ADVANCES IN CHEMICAL ENGINEERING

Volume 15

ADVANCES IN
CHEMICAL ENGINEERING

Editor-in-Chief
JAMES WEI
*Department of Chemical Engineering
Massachusetts Institute of Technology
Cambridge, Massachusetts*

Editors

JOHN L. ANDERSON
*Department of Chemical Engineering
Carnegie-Mellon University
Pittsburgh, Pennsylvania*

KENNETH B. BISCHOFF
*Department of Chemical Engineering
University of Delaware
Newark, Delaware*

JOHN H. SEINFELD
*Department of Chemical Engineering
California Institute of Technology
Pasadena, California*

Volume 15

ACADEMIC PRESS, INC.
Harcourt Brace Jovanovich, Publishers
San Diego New York Boston
London Sydney Tokyo Toronto

TP
145
.D7
1990
v.15

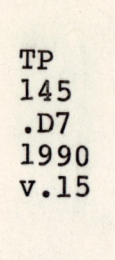

This book is printed on acid-free paper. ∞

COPYRIGHT © 1990 BY ACADEMIC PRESS, INC.
All Rights Reserved.
No part of this publication may be reproduced or transmitted in any form or by any means, electronic or mechanical, including photocopy, recording, or any information storage and retrieval system, without permission in writing from the publisher.

ACADEMIC PRESS, INC.
San Diego, California 92101

United Kingdom Edition published by
ACADEMIC PRESS LIMITED
24-28 Oval Road, London NW1 7DX

LIBRARY OF CONGRESS CATALOG CARD NUMBER: 56-6600

ISBN 0-12-008515-1 (alk. paper)

PRINTED IN THE UNITED STATES OF AMERICA
90 91 92 93   9 8 7 6 5 4 3 2 1

CONTENTS

PREFACE vii

Rheological Models of Suspensions
Pierre M. Adler, Ali Nadim, and Howard Brenner

I.	Introduction	1
II.	General Remarks	6
III.	Empirical and Cell Models	19
IV.	Dilute Suspensions	23
V.	Statistical Models	28
VI.	Percolation Models	32
VII.	Spatially Periodic Suspensions	36
VIII.	Current Research Topics	54
	References	67

Opportunities in the Design of Inherently Safer Chemical Plants
Stanley M. Englund

I.	Introduction	73
II.	Identification of Hazards	74
III.	General Design Opportunities	75
IV.	Process Design Opportunities	85
V.	Equipment Design Opportunities	104
VI.	Conclusion	133
	References	134

Interactions between Colloidal Particles and Soluble Polymers
H. J. Ploehn and W. B. Russel

I.	Introduction	137
II.	Polymer Solution Thermodynamics	140
III.	Randomly Absorbing Homopolymers	157
IV.	Terminally Anchored Polymers	197
V.	Nonadsorbing Polymer	205
VI.	Macroscopic Consequences	211
	References	224

INDEX 229

PREFACE

The Frontier Report of the National Research Council analyzed chemical engineering activities on three scales: the microscale of atomic and molecular events, the mesoscale of process equipment, and the macroscale of entire systems. The traditional home of chemical engineers has been in the mesoscale of unit operations and reaction engineering and, more recently, in the microscale of transport phenomena and chemical kinetics. Increasingly, our attention is drawn to macroscale problems: chemical plant safety, environmental impact of hazardous chemicals and even carbon dioxide, manufacturing productivity and world competition, and the introduction of products with innovative properties. The chemical engineers have to address the "big picture," or someone else who does not have the knowledge and expertise of chemical processing will make decisions that may be costly and unwise or even counterproductive.

There are three chapters in this volume, two of which address the microscale. Ploehn and Russel address the "Interactions Between Colloidal Particles and Soluble Polymers," which is motivated by advances in statistical mechanics and scaling theories, as well as by the importance of numerous polymeric flocculants, dispersants, surfactants, and thickeners. How do polymers thicken ketchup? Adler, Nadim, and Brenner address "Rheological Models of Suspensions," a closely related subject through fluid mechanics, statistical physics, and continuum theory. Their work is also inspired by industrial processes such as paint, pulp and paper, and concrete; and by natural systems such as blood flow and the transportation of sediment in oceans and rivers. Why did doctors in the Middle Ages induce bleeding in their patients in order to thin their blood?

The remaining chapter is in the macroscale, where Englund addresses "Opportunities in the Design of Inherently Safer Chemical Plants." He carefully points out that there is no such thing as an *inherently safe* chemical plant, since there is always risk, but we can make chemical plants *inherently safer*. The major and minor chemical plant accidents in the past years are constant reminders of the importance of this topic. The day will come when this topic will move from industrial good practice to university research, then to undergraduate electives, and eventually to mandatory undergraduate education. If we do a good job, we will earn the respect of the public, who will then become less suspicious of our motives and methods.

James Wei

RHEOLOGICAL MODELS OF SUSPENSIONS

Pierre M. Adler

*Laboratoire d'Aérothermique
Centre National de la Recherche
F-92190 Meudon, France*

Ali Nadim

*Department of Mathematics
Massachusetts Institute of Technology
Cambridge, Massachusetts 02139*

Howard Brenner

*Department of Chemical Engineering
Massachusetts Institute of Technology
Cambridge, Massachusetts 02139*

I. Introduction

Determination of the rheological and other dynamical properties of suspensions constitutes a field of fundamental importance to engineers, applied physicists, chemists, biologists, and industrial practitioners. From a practical standpoint, information on the macroscopic behavior of suspensions owes its importance to the many industrial processes that require the movement and processing of suspensions. Obvious examples occur in heavy industries such as the paint, pulp and paper, and concrete industries. Ferrofluids, for example, constitute a high technology suspension and possess novel properties useful in specialized industrial applications. Suspensions also occur naturally; examples are blood flow and the transportation of sediment in oceans and rivers. On the less pragmatic side, suspensions are of much theoretical interest. They provide an area of fundamental research, combining elements of fluid mechanics, statistical physics, and continuum theory, that focuses on the goal of describing the macroscopic properties of such "heterogeneous continua" starting with the exact laws that govern the interstitial fluid and particulate-phase properties at the microscale.

Until the early 1980's, studies of suspensions were largely confined to the engineering sciences. However, the interdisciplinary nature of suspension rheology has been recognized, and now, many significant contributions are being made by chemists and physicists as well as engineers and applied mathematicians. This chapter is intended to collect existing models of suspensions together within a unified framework that aims to bridge parochial disciplinary fragmentation of the subject.

Many review articles on suspension theories already exist. Some are listed chronologically in Table I together with a brief commentary attempting to capture the primary focus of each. Among other things, such a categorical listing provides a brief historical account of the growth of this technologically important field since the late 1960s. It also furnishes some insight into the more specialized subdivisions into which suspension theory has branched. The present review will focus mainly on theoretical results pertaining to the rheological properties of multiparticle suspensions. Single particle (i.e., dilute) results have been sufficiently reviewed in the past (cf. Table I for references) to render unnecessary detailed commentary here. Moreover, experimental results will not generally be reviewed, although some will be cited at appropriate junctures to provide comparison with existing theories.

Since this chapter is concerned mainly with reviewing existing models of suspensions, such complicating factors as nonNewtonian suspending fluids, inertia, and wall (i.e., noncontinuum) effects will be excluded for the most part. Neglect of inertial effects is tantamount to requiring the translational, rotary, and shear Reynolds numbers (based on particle size) to be small. In such circumstances, and for incompressible Newtonian suspending fluids, the microscale motion of the interstitial fluid is governed by the Stokes and continuity equations, each of which is linear. The latter property renders tractable the actual solutions of boundary-value problems and confers a convenient linear form upon important global quantities such as hydrodynamic forces and torques, as outlined in the next section. On the other hand (in the absence of Brownian motion), neglect of inertia may render indeterminate (Leal, 1980) the Stokes-flow problem thereby posed. Examples of this indeterminacy are provided by the lateral position of a neutrally buoyant sphere relative to the wall bounding a simple shear flow, or the terminal orientation of a non-spherical particle sedimenting in a quiescent fluid. Such indeterminacy has interesting implications regarding the often assumed ergodicity of purely mechanical multiphase transport processes. As a rule, the absence of inertia favors nonergodic states, whereas inclusion of inertial, nonNewtonian, and/or Brownian effects removes the configurational indeterminacy of the Stokes equations (Leal, 1979, 1980).

TABLE I

Major Reviews of Suspension Theories

Author	Content
Happel and Brenner (1965)	A theoretical overview; a comprehensive review of the literature prior to 1965.
Goldsmith and Mason 1967)	An account of experimental observations.
Brenner (1970)	General definitions and relations for the rheological properties of two-phase systems, mainly dilute.
Cox and Mason (1971)	Motions of single particles.
Brenner (1972a)	Suspension rheology; rigid and deformable particles.
Brenner (1972b)	Particle dynamics; wall effects; inertial effects; inhomogeneous shear fields; Brownian motion.
Sather and Lee (1972)	A statistical approach.
Batchelor (1974)	General properties of random two-phase materials; analogies; bounds, dilute suspensions at order ϕ and ϕ^2.
Brenner (1974)	An exhaustive account of the rheology of dilute suspensions of axisymmetric Brownian particles.
Jinescu (1974)	Classification of the rheological behavior of suspensions.
Batchelor (1976a)	A review of consequences of various forces acting on suspensions; dilute suspensions at order ϕ and ϕ^2.
Jeffrey and Acrivos (1976)	A discussion of major experimental and theoretical works.
Mewis and Spaull (1976)	Rheological behavior; main forces involved; structure.
Mason (1977)	An account of the work of Mason and co-workers.
Saville (1977)	Movement of ionic solutions near charged interfaces.
Buyevich and Shchelchkova (1978)	A statistical approach; phase averages.
Leal (1979)	NonNewtonian suspending fluids; cumulative effects of weak and strong viscoelasticity.
Brunn (1980)	NonNewtonian suspending fluids; their influence on the motion of an isolated particle and on particle–particle and particle–wall interactions.
Herczyński and Pieńkowska (1980)	A statistical approach; macroscopic equations; hierarchy closure.
Leal (1980)	The indeterminate character of some Stokes flow problems; influence of a nonNewtonian suspending fluid, inertia and deformability.
Mewis (1980)	An account of rheological behavior; flocculated suspensions.
Russel (1980)	Effects of colloidal forces (i.e. van der Waals, electrostatic, and Brownian forces).
Davis and Acrivos (1985)	Sedimentation in monodisperse and polydisperse suspensions; effect of wall inclination.
Feuillebois (1988)	Particle–particle and particle–wall interactions; inertial effects; unsteadiness.
Metzner (1985)	Suspensions in polymeric liquids.
Barnes et al., (1987)	Applications of computer simulations to dense suspension rheology.
Bird et al., (1987)	Suspensions of macromolecules from a statistical mechanical viewpoint.
Brady and Bossis (1988)	Molecular-dynamics-like simulation of suspensions ("Stokesian dynamics")

This chapter will focus on infinitely-extended suspensions in which potential complications introduced by the presence of walls are avoided. The only wall-effect case that can be treated with relative ease is the interaction of a sphere with a plane wall (Goldman et al., 1967a,b). The presence of walls can lead to relevant suspension rheological effects (Tözeren and Skalak, 1977; Brunn, 1981), which result from the existence of particle depletion boundary layers (Cox and Brenner, 1971) in the proximity of the walls arising from the finite size of the suspended spheres. Going beyond the dilute and semidilute regions considered by the authors just mentioned is the ad hoc percolation approach, in which an infinite cluster—assumed to occur above some threshold particle concentration—necessarily interacts with the walls (cf. Section VI).

Other complicating effects arise from the presence of gravity, interparticle forces, and Brownian motion. Gravity (or some equivalent external force field) constitutes the driving force for sedimentation processes. Interparticle forces such as double-layer forces, important when the suspended particles are both small and charged (and the suspending fluid is an electrolyte), as well as London-van der Waals attractive forces, also important for small particles, have attracted the attention of colloid scientists for many years. Saville (1977) and Russel (1980) review many aspects of such colloidal forces in the context of suspensions. The influence of a macroscopic electric field on conducting particles in suspension may also be important. [See, e.g., Arp and Mason (1977a), and Adler (1981b) for two-sphere interactions.] However, such effects will be excluded from this chapter; a discussion of aggregable suspensions, which requires consideration of interparticle forces, will also be omitted, but since the latter field has important fundamental and practical prospects, a few contributions will be cited. Adler and Mills (1979) and Adler (1979) analyzed the rupture by hydrodynamic forces of particulate aggregates undergoing both simple shear and sedimentation. Subsequently, this topic was extended to fractal flocs by Sonntag and Russel (1986). van de Ven and Hunter (1979) introduced an elastic floc model that displayed general agreement with oscillating flow experiments, and van Diemen and Stein (1983) conducted experiments with aqueous quartz dispersions.

Brownian motion must be taken into account for suspensions of small (submicron-sized) particles. By their very nature, such stochastic Brownian forces favor the ergodicity of any configurational state. Although no completely general framework for the inclusion of Brownian motion will be presented here, its effects will be incorporated within specific contexts. Especially relevant, in the present rheological context, is the recent review by Felderhof (1988) of the contribution of Brownian motion to the viscosity of suspensions of spherical particles.

This chapter is organized as follows. Section II provides a general introduction to the theoretical models. Starting with the equations of motion, valid locally, the grand-resistance matrix formulation of this multiparticle system is introduced. Various specific solutions of the many-body problem in low Reynolds-number hydrodynamics are discussed, followed by a statistical formulation of this multiparticle system. Empirical and so-called cell models are discussed in Section III. Historically, cell models were among the first to be employed to account for particle–particle hydrodynamic interactions. They have since proven valuable for correlation purposes and order-of-magnitude phenomenological coefficient estimates, though they lack a rational geometric basis. Section IV addresses dilute suspensions in which the particulate volumetric fraction ϕ is assumed sufficiently small such that only one-, or at most, two-body particle interactions with the fluid are important. Statistical models of suspensions are reviewed in Section V. Formal expansions are first described together with a number of proposed closure schemes for the resulting hierarchy of equations. Monte Carlo calculations, both with and without hydrodynamic interactions among particles, are briefly surveyed.

Section VI outlines percolation-theory approaches to the modeling of suspensions. Following a general introduction to percolation concepts, some theoretical and experimental developments are described. Above some threshold concentration, an infinite cluster of particles forms as a result of the nonzero time interval over which adjacent particles sensibly interact hydrodynamically. Section VII reviews contributions to the rheology of ordered models of suspensions. When a single particle is contained within the unit cell of a spatially periodic suspension, the kinematics of the flow are known a priori, at which point the dynamical problem can be solved completely. Novel concepts such as maximum kinematic concentration, self-reproducibility of the lattice in time, and time-averaged rheological properties are discussed. Finally, four "current research" areas are reviewed in Section VIII. The first, Stokesian dynamics, consists of a molecular-dynamics-type simulation technique applied to suspensions, highlighting hydrodynamic interactions among the suspended particles. The second is the application of generalized Taylor dispersion theory to the analysis of momentum transport processes in suspensions. (Included in this category is a discussion of the significant experimental contributions of Mondy, Graham and co-workers to the falling-ball viscometry of suspensions.) The third area addresses suspensions possessing fractal structure, which are important for concentrated or heterogeneous suspensions. Lastly, some novel aspects of magnetic fluid phenomena are discussed.

Notation and Scope

To facilitate references to mathematical equations appearing in the original literature, we have in (almost) all cases retained the author's original notation in preparing our review rather than attempting a premature standardization. While this results in minor notational inconsistencies in the text [e.g., the angular velocity of a rigid rotating particle is represented by ω in Eq. (2.4) but by Ω in Eq. (7.11)], we believe any momentary confusion resulting from our convention will be more than offset by ease of access to the author's original equations.

The scope of this chapter is admittedly subjective, reflecting to a large extent the current individual and collective interests of its authors. Accordingly, no attempt is made to cover *all* technologically and scientifically important aspects of suspension rheology. For instance, our focus is geared more towards suspensions in general than towards the specific properties of, say, suspensions of colloidal and molecular particulates. As such, although some discussion of colloidal forces, Brownian motion, and "flexible" bodies appears, our review does not, in any sense, provide a comprehensive survey of so-called "colloidal" suspensions. The interested reader will find here, *inter alia*, an extended overview of spatially periodic and fractal models of suspensions as well as a description of momentum tracer schemes for probing suspension rheological properties—a new unsteady-state scheme wholly different from classical rheometric methods. In a periodical series of books bearing the name "*Advances in...*," such subjective choices of topics is both inevitable and, we hope, desirable.

II. General Remarks

This section furnishes a brief overview of the general formulation of the hydrodynamics of suspensions. Basic kinematical and dynamical microscale equations are presented, and their main attributes are described. Solutions of the many-body problem in low Reynolds-number flows are then briefly exposed. Finally, the microscale equations are embedded in a statistical framework, and relevant volume and surface averages are defined, which is a prerequisite to describing the macroscale properties of the suspension.

A. Basic Equations and Properties

Consider an unbounded, incompressible, Newtonian fluid undergoing a homogeneous shear flow characterized by the position-independent velocity gradient dyadic **G**, which can be decomposed into symmetric and antisymmetric contributions **S** and $\ddot{\Lambda}$, respectively, as

$$\mathbf{G} = \mathbf{S} + \ddot{\Lambda}. \tag{2.1}$$

Let μ_0 denote the fluid viscosity. Consider N suspended particles, not necessarily identical, enumerated by the index $i = 1, 2, \ldots, N$. At each instant of time, the spatial and orientational geometric configuration of the particles, each assumed rigid, is completely determined upon specification of $3N$ independent spatial, scalar coordinates of designated locator points affixed to each particle, and $3N$ orientational coordinates (e.g., the three Eulerian angles of each of the N particles) specifying their orientations. Allowing the 3-vectors $\mathbf{x}_i$ and (symbolically) $\mathbf{e}_i$ to denote the respective spatial and orientational coordinates of particle i, one can collectively denote the configurational coordinates of all the particles as $(\mathbf{x}^N, \mathbf{e}^N)$. The values of the spatial and orientational variables $\mathbf{x}^N$ and $\mathbf{e}^N$ together define a point in a $6N$-dimensional space, the phase space of the particulate phase, symbolized by Γ_N (Hansen and McDonald, 1976).

When all relevant Reynolds numbers are sufficiently small to permit neglect of inertial effects, the interstitial fluid velocity and pressure fields $\mathbf{v}$ and p, respectively, satisfy the quasistatic Stokes and continuity equations,

$$\mu_0 \nabla^2 \mathbf{v} = \nabla p, \tag{2.2}$$

$$\nabla \cdot \mathbf{v} = 0. \tag{2.3}$$

Denote by 0_i the locator point of i, and denote by $\mathbf{r}_i$ a position vector drawn relative to 0_i. Suppose 0_i to move with velocity $\mathbf{U}_i$ relative to a space-fixed coordinate system, and let the particle possess angular velocity $\boldsymbol{\omega}_i \equiv d\mathbf{e}_i/dt$ relative to the latter system. The no-slip condition on the surface s_i of particle i then takes the form

$$\mathbf{v} = \mathbf{U}_i + \boldsymbol{\omega}_i \times \mathbf{r}_i \quad \text{on } s_i \quad (i = 1, \ldots, N). \tag{2.4}$$

Equations (2.2) and (2.3) are to be solved subject to Eq. (2.4), together with a condition imposed on the "average" velocity field (discussed in the following paragraphs) in order to obtain the fluid velocity and pressure fields. Furthermore, the translational and angular particle velocities $\mathbf{U}_i$ and $\boldsymbol{\omega}_i$ appearing in Eq. (2.4) are usually not specified a priori; more often, it is the external forces and torques acting on the particles that are assumed known. In such circumstances, force- and torque-free conditions imposed on the particles (consistent with neglect of particle inertia) are employed to calculate the velocities $\mathbf{U}_i$ and $\boldsymbol{\omega}_i$.

When N is finite, although nevertheless possibly large, the particles are necessarily confined externally within a finite volume of the infinitely extended fluid. The velocity field must then asymptotically approach, far from the outermost particles, the undisturbed velocity field corresponding to $\mathbf{G}$; explicitly,

$$\mathbf{v} \to \mathbf{v}_\infty, \tag{2.5a}$$

in which

$$\mathbf{v}_\infty = \mathbf{U}_0 + \boldsymbol{\omega}_f \times \mathbf{r}_0 + \mathbf{S} \cdot \mathbf{r}_0, \tag{2.5b}$$

with $\mathbf{U}_0$ the translational velocity vector of the undisturbed fluid at a conveniently chosen reference point 0, $\mathbf{r}_0$ the position vector relative to 0, and $\omega_f = -\frac{1}{2}\ddot{\varepsilon}:\ddot{\Lambda} \equiv \frac{1}{2}\nabla \times \mathbf{v}_\infty$, which is half the undisturbed vorticity vector ($\ddot{\varepsilon} \equiv \varepsilon_{ijk}$, the unit isotropic triadic).

When N is infinite, so that suspended particles are dispersed throughout the entire fluid domain, condition (2.5) is replaced by an equivalent one that prescribes the "average" fields $\mathbf{U}_0$ and $\mathbf{G}$. These are explicitly given for spatially periodic suspensions in Section VII. [See also the paragraph following Eq. (2.28).]

The detailed solution of Eqs. (2.2) and (2.3), subject to the given boundary and/or mean-field conditions, represents a formidable problem and has been solved only for a small number N of particles. However, making use of the linear character of the governing equations, much can nonetheless be formally concluded regarding the functional dependence of the solution. In such context, we consider the formalism of Brenner and O'Neill (1972) in the following paragraph, albeit with minor modifications.

The hydrodynamic force $\mathbf{F}_i$ and torque $\mathbf{T}_i$ (about 0_i) exerted by the fluid on particle i are given respectively by the expressions

$$\mathbf{F}_i = \int_{s_i} d\mathbf{s} \cdot \mathbf{P}, \tag{2.6a}$$

$$\mathbf{T}_i = \int_{s_i} \mathbf{r}_i \times (d\mathbf{s} \cdot \mathbf{P}), \tag{2.6b}$$

in which $d\mathbf{s}$ is the outwardly-directed vector surface-element on the particle surface s_i pointing into the fluid, and

$$\mathbf{P} = -\mathbf{I}p + \mu_0(\nabla \mathbf{v} + \nabla \mathbf{v}^\dagger) \tag{2.7}$$

is the stress tensor for the Newtonian fluid.

The particle stress $\mathbf{A}$, defined as the contribution to the bulk stress arising from the presence of the suspended particles in the fluid (Batchelor, 1970), possesses the symmetric and traceless deviatoric form

$$\mathbf{A} = \frac{1}{V} \sum_{i=1}^{N} \ddot{\mathscr{S}}_i, \tag{2.8a}$$

where

$$\ddot{\mathscr{S}}_i = \frac{1}{2} \int_{s_i} [d\mathbf{s} \cdot \mathbf{P}\mathbf{r}_i + \mathbf{r}_i d\mathbf{s} \cdot \mathbf{P} - \tfrac{2}{3}\mathbf{I}(d\mathbf{s} \cdot \mathbf{P} \cdot \mathbf{r}_i)]. \tag{2.8b}$$

Owing to the underlying linearity of the problem (including the assumed rigidity of the particles), the forces, torques, and particle stress are each necessarily linear functions of the kinematical quantities $U_i - u_i$, $\omega_i - \omega_f$ and S (where u_i denotes the velocity of the undisturbed flow v_∞ existing at point 0_i in the absence of particles). These linear relations may be written as the partitioned matrix relation

$$\begin{bmatrix} F^N \\ T^N \\ A \end{bmatrix} = \mu_0 R(x^N, e^N) \cdot \begin{bmatrix} U^N - u^N \\ \omega^N - \omega_f^N \\ S \end{bmatrix}. \tag{2.9}$$

Notation similar to that introduced previously for (x^N, e^N) has been used to collectively denote the forces F^N, torques T^N, velocities U^N, etc., of all the N particles. Relation (2.9) is similar to one of Brenner and O'Neill (1972) together with Hinch's (1972) adjunction of A.

As a consequence of Lorentz' reciprocal theorem (see Happel and Brenner, 1965) the grand resistance matrix $R(x^N, e^N)$ possesses many internal symmetries, greatly reducing the number of its independent elements. Another important feature of R is that it depends only on the instantaneous configuration (x^N, e^N) of the particulate phase.

Denote by $F_{i, \text{ext}}$ and $T_{i, \text{ext}}$ the external force and torque (about 0_i) acting on particle i. Neglect of particle inertia then leads to the equation

$$\begin{bmatrix} F^N \\ T^N \end{bmatrix} + \begin{bmatrix} F^N_{\text{ext}} \\ T^N_{\text{ext}} \end{bmatrix} = 0, \tag{2.10}$$

governing the instantaneous particle dynamics. Substitution of Eq. (2.9) into Eq. (2.10) yields a first-order system of differential equations governing the trajectory of the point (x^N, e^N) in the phase space Γ_N. Consider, for instance, a suspension of N force- and couple-free spheres. When the latter suspension is sheared, Eqs. (2.9) and (2.10) yield an equation of the structural form

$$dx^N/dt = f(x^N), \tag{2.11}$$

with f as a nonlinear function of the particle positions (functionally dependent also on S); its form can be deduced from the grand resistance matrix R.

The structure of the system in Eq. (2.11) is formally very simple, although apart from the kinematic reversibility of the individual particle motions, which is a consequence of the time invariance of the quasistatic Stokes and continuity equations (Slattery, 1964), very little else can be said explicitly. Equation (2.11) would appear to pose a fruitful future study within the more general framework of dynamical systems (Collet and Eckmann, 1980) whose temporal evolution is governed by a system of equations identical in structure to Eq. (2.11). Although this interesting new field is still in its infancy (Aref and Balachandar, 1986), it appears that widely differing behaviors (i.e., 'chaos') may be exhibited by otherwise identical systems that differ only insignificantly in their initial configurations.

In this context, consider, for example, several existing results pertaining to the temporal behavior of multiparticle systems. Two-sphere systems are the only ones completely analyzed to date. In simple shear, the relative trajectory of two spheres may be either open or closed, depending on their initial separation (Batchelor and Green, 1972a); thus, the corresponding temporal behavior is either aperiodic or periodic. In elongational flows, the relative trajectories are always open. Other examples are provided by sedimenting spheres. Two identical spheres settle at the same velocity (Goldman et al., 1966), thus maintaining their original configuration, whereas two unequal spheres settle at different speeds. Three or more equal, vertically-aligned spheres exhibit the phenomenon of "critical initial spacing" (Leichtberg et al., 1976). Depending on the initial spacing, the gap between the two leading spheres in the chain may asymptote in time, either towards zero or to some finite separation distance. Ganatos et al. (1978) have further examined three-sphere configurations.

B. Solutions of the Many-Body Problem in Low Reynolds-Number Hydrodynamics

Having exposed the general fluid-mechanical and particulate equations, attention is now directed towards reviewing the many-body problem as governed by Eqs. (2.2)–(2.5). Unless otherwise stated, our focus is exclusively on rigid-sphere suspensions. Historically, the earliest study is that of G. G. Stokes (1851), who calculated the drag force on a sphere translating through an otherwise quiescent viscous fluid. For the sake of completeness, and as one of the few analytical results that will be cited here, we also recall that the velocity and pressure fields about a sphere of radius a suspended in a sheared fluid with the velocity at infinity given by Eq. (2.5) are, respectively

$$\mathbf{v} = \mathbf{U}_0 + (\mathbf{U} - \mathbf{U}_0) \cdot \left[\frac{3}{4}\frac{a}{r}(\mathbf{I} + \hat{\mathbf{r}}\hat{\mathbf{r}}) + \frac{1}{4}\left(\frac{a}{r}\right)^3 (\mathbf{I} - 3\hat{\mathbf{r}}\hat{\mathbf{r}}) \right]$$
$$+ \omega_f \times \mathbf{r} + \left(\frac{a}{r}\right)^3 (\omega - \omega_f) \times \mathbf{r}$$
$$+ \mathbf{S} \cdot \mathbf{r}\left[1 - \left(\frac{a}{r}\right)^5\right] - \frac{5}{2}\left(\frac{a}{r}\right)^3 \left[1 - \left(\frac{a}{r}\right)^2\right] r\mathbf{S}:\hat{\mathbf{r}}\hat{\mathbf{r}}\hat{\mathbf{r}} \quad (2.12a)$$

and

$$p - p_\infty = \frac{3\mu_0 a}{2r^2}\hat{\mathbf{r}}\cdot(\mathbf{U} - \mathbf{U}_0) - 5\mu_0\left(\frac{a}{r}\right)^3 \mathbf{S}:\hat{\mathbf{r}}\hat{\mathbf{r}}. \qquad (2.12b)$$

Here, the sphere center is instantaneously situated at point 0; the sphere center translates with velocity $\mathbf{U}$, while it rotates with angular velocity ω; $\mathbf{r}$ is measured relative to 0; its magnitude $|\mathbf{r}|$ is denoted by r. Moreover, $\hat{\mathbf{r}} = \mathbf{r}/r$ is a unit radial vector. The latter solution is derivable in a variety of ways; e.g., from Lamb's (1932) general solution (Brenner, 1970). [Equation (2.12) represents a superposition (Brenner, 1958) of three physically distinct solutions, corresponding, respectively, to (i) translation of a sphere through a fluid at rest at infinity; (ii) rotation of a sphere in a fluid at rest at infinity; (iii) motion of a neutrally buoyant sphere suspended in a linear shear flow. The latter was first obtained by Einstein (1906, 1911; cf. Einstein, 1956) in connection with his classic calculation of the viscosity of a dilute suspension of spheres, which formed part of his 1905 Ph.D. thesis.]

Passing beyond the isolated sphere case, two spheres pose the simplest "interaction" problem arising in low Reynolds-number flows. Studies of the these originated years ago (Stimson and Jeffery, 1926; Manley and Mason, 1952), although complete and systematic results for all separation distances and relative sphere sizes were obtained only much later. For equal spheres, these results were given by Batchelor and Green (1972a), Arp and Mason (1977b), Kim and Mifflin (1985), Yoon and Kim (1987), and Cichocki et al. (1988). Results for unequal spheres were tabulated by Adler (1981a) and, in a more complete form, by Jeffrey and Onishi (1984). Such two-sphere interaction calculations are important not only because they are directly relevant to dilute suspension analyses, but their existence also provides a criterion of the accuracy of numerical or computational schemes for the more general N-particle case.

In a series of papers, Felderhof has devised various methods to solve anew one- and two-sphere Stokes flow problems. First, the classical "method of reflections" (Happel and Brenner, 1965) was modified and employed to examine two-sphere interactions with mixed slip-stick boundary conditions (Felderhof, 1977; Renland et al., 1978). A novel feature of the latter approach is the use of "superposition" of forces rather than of velocities; as such, the mobility matrix (rather than its inverse, the grand resistance matrix) was derived. Calculations based thereon proved easier, and convergence was more rapid; explicit results through terms of $0(\rho^{-7})$ were derived, where ρ is the nondimensional center-to-center distance between spheres. In a related work, Schmitz and Felderhof (1978) solved Stokes equations around a sphere by the so-called Cartesian ansatz method, avoiding the use of spherical coordinates. They also devised a second method (Schmitz and Felderhof, 1982a), in which this Cartesian problem is formulated as an expansion of scattered waves. Through use of the same scheme, as previously mentioned, this method was extended to include two-sphere interactions (Schmitz and Felderhof, 1982b,c) and obtained terms of $0(\rho^{-12})$. Efforts are underway (Kim, 1986) to extend these calculations to the interaction of ellipsoidal particles.

None of these two-body methods has yet been extended to the many-body problem. A contribution by Kynch (1959) applied a reflection scheme to the three-body problem, although explicit details were not provided.

A suspension composed of an infinitely-extended regular array of identical spheres was examined by Kapral and Bedeaux (1978) as an application of the scheme developed by Bedeaux et al. (1977) [and independently by Zuzovsky et al. (1983)] with similar results. This problem will be encountered again in Section VII. In a certain sense, the multiparticle problem is thereby reduced to a single particle problem (within a unit cell-boundary), since all of the particles exist in identical states. That is, each is kinematically and dynamically indistinguishable from the others.

Mazur (1982) and Mazur and van Saarloos (1982) developed the so-called method of "induced forces" in order to examine hydrodynamic interactions among many spheres. These forces are expanded in irreducible induced-force multipoles and in a hierarchy of equations obtained for these multipoles when the boundary conditions on each sphere were employed. Mobilities are subsequently derived as a power series-expansion in ρ^{-1}. In principle, calculations may be performed to any order, having been carried out by the above authors through terms of $0(\rho^{-7})$ for a suspension in a quiescent fluid. To that order, hydrodynamic interactions between two, three, and four spheres all contribute to the final result. This work is reviewed by Mazur (1987).

A numerical solution technique was developed by Leichtberg et al. (1976) and Ganatos et al. (1978) to investigate the coaxial settling of three or more spheres, as well as the motion of identical spheres in a horizontal plane; the fluid is assumed quiescent in all cases. A collocation technique was used in the latter problem to determine the coefficients appearing in the expansion of the general solution of Stokes equations by demanding the boundary conditions to be satisfied at a finite number of points lying on the sphere surfaces. The choice of these points is a delicate problem. No systematic extension of this method has yet been made. A recent analysis (Hassonjee et al., 1988) extends the original Ganatos et al. (1978) treatment to asymmetric clusters of spherical particles.

In conclusion, much remains to be done in the field of many-body hydrodynamic interactions. Existing results need to be embedded into a unified and systematic framework and extended from quiescent to sheared suspensions. The methods of Mazur and co-workers can be used to derive far-field approximations, numerical techniques (e.g., collocation) can supply the middle range, and "lubrication" approximations (see Section VII) can furnish limiting asymptotic results for two or more spheres in close proximity to one another.

C. Statistical Formulation

This subsection attempts a statistical description of the multiparticle problem. Such developments closely follow those employed in the classical theory of liquids (Hansen and McDonald, 1976), together with the contribution of Batchelor (1970), as well as certain technical elements reviewed by Herczyński and Pieńkowska (1980).

Consider a statistically homogeneous dispersion of N particles contained in a volume V. Although generalization to the case of dissimiliar particles is not difficult, the particles are assumed for simplicity to be identical spheres (radii a). For any given experimental realization, the configuration of the latter particulate system at time t is completely determined by the values of the $3N$ coordinates $\mathbf{x}^N$ representing the particle locator points (chosen to lie at the sphere centers so as to preclude individual particle orientational issues) of the N spheres. Note that the dynamical variables (i.e., momenta) need not be included in the phase space since inertia has been systematically neglected.

Suppose the particles are distinguished by numbering them from 1 to N; denote by

$$F^{(N)}(\mathbf{x}^N, t)\, d\mathbf{x}^N \qquad (2.13)$$

the probability that at time t in a given realization (the locator point for) particle 1 lies within the interval $(\mathbf{x}_1, \mathbf{x}_1 + d\mathbf{x}_1)$ while, simultaneously, particle 2 is located within the region $(\mathbf{x}_2, \mathbf{x}_2 + d\mathbf{x}_2)$, etc. In Eq. (2.13),

$$d\mathbf{x}^N = \prod_{i=1}^{N} d\mathbf{x}_i$$

constitutes the phase–space volume element centered at $\mathbf{x}^N$. The particle-specific probability density function $F^{(N)}$ represents the limiting value of the ratio of the number of times that such a configuration obtains to the total number of realizations in the limit as the latter number approaches infinity.

It is important to recognize that major theoretical difficulties arise in the limiting process $N \to \infty$. This highlights the important differences existing between the cases of finite (although very large) N and infinite N. These issues are addressed in Section VII,D.

A series of lower-order particle-specific probability densities $F^{(n)}$ ($n \leq N$) may be defined as

$$F^{(n)}(\mathbf{x}^n, t) = \int \cdots \int F^{(N)}(\mathbf{x}^N, t)\, d\mathbf{x}_{n+1}\, d\mathbf{x}_{n+2}, \ldots, d\mathbf{x}_N, \qquad (2.14)$$

possessing a similar interpretation to that of Eq. (2.13) for the first n particles, irrespective of where the remaining $N-n$ particles may be located. Probability densities in Eqs. (2.13) and (2.14) are to be regarded as normalized in the sense that

$$\int F^{(N)}(\mathbf{x}^N, t)\, d\mathbf{x}^N = 1. \qquad (2.15)$$

Requisite integrations appearing in Eqs. (2.14) and (2.15) are to be performed over the entire volume available to the particles for each $d\mathbf{x}_i$. Given the indistinguishability of the (identical) particles, it is often more convenient to define the (generic) probability density $P^N(\mathbf{x}^N, t)$ that *any* particle is located in $(\mathbf{x}_1, \mathbf{x}_1 + d\mathbf{x}_1)$, any other in $(\mathbf{x}_2, \mathbf{x}_2 + d\mathbf{x}_2)$, etc. It is easily verified that

$$P^{(N)}(\mathbf{x}^N, t) = N!\, F^{(N)}(\mathbf{x}^N, t). \qquad (2.16)$$

Similarly, the lower-order generic density $P^{(n)}(\mathbf{x}^n, t)$ can be demonstrated to be related to its particle-specific counterpart by the expression

$$P^{(n)}(\mathbf{x}^n, t) = \frac{N!}{(N-n)!} F^{(n)}(\mathbf{x}^n, t), \qquad (2.17)$$

which reduces to Eq. (2.16) when $n = N$. Note the normalization requirement imposed on the latter generic density requires that

$$\int \cdots \int P^{(n)}(\mathbf{x}^n, t)\, d\mathbf{x}_1, \ldots, d\mathbf{x}_n = N!/(N-n)!, \qquad (2.18)$$

in which the numerical factor must be accounted for in calculating the expected values of any configuration-dependent quantity.

Of particular importance in most theories are the first two densities $P^{(1)}$ and $P^{(2)}$. The first, $P^{(1)}(\mathbf{x}, t)$ represents the probability density for finding a particle at $\mathbf{x}$ at time t; it is equal to the average volumetric particulate number density n:

$$P^{(1)}(\mathbf{x}, t) = n. \qquad (2.19)$$

For a stationary homogeneous system, n is a constant given by N/V. (Note that n here is not a counting index.) $P^{(2)}(\mathbf{x}_1, \mathbf{x}_2, t)$ represents the probability density for finding two particles at $\mathbf{x}_1$ and $\mathbf{x}_2$ simultaneously. For a homogeneous system it depends on the difference $\mathbf{x}_1 - \mathbf{x}_2$ rather than on $\mathbf{x}_1$ and $\mathbf{x}_2$

individually:

$$P^{(2)} \equiv P^{(2)}(\mathbf{x}_1 - \mathbf{x}_2, t). \tag{2.20}$$

Also of importance is the conditional probability density $F_c^{(N-1)}$ for finding particles $2-N$, respectively, at $\mathbf{x}_2-\mathbf{x}_N$, given that particle 1 is located at $\mathbf{x}_1$. The relation

$$F^{(N)}(\mathbf{x}_1, \ldots, \mathbf{x}_N, t) = F^{(1)}(\mathbf{x}_1, t) F_c^{(N-1)}(\mathbf{x}_2, \ldots, \mathbf{x}_N, t | \mathbf{x}_1), \tag{2.21}$$

governing the latter conditional probability, may be written for its generic counterpart $P_c^{(N-1)}$ as

$$P^{(N)}(\mathbf{x}_1, \ldots, \mathbf{x}_N, t) = P^{(1)}(\mathbf{x}_1, t) P_c^{(N-1)}(\mathbf{x}_2, \ldots, \mathbf{x}_N, t | \mathbf{x}_1) \tag{2.22}$$

upon multiplication of Eq. (2.21) by $N!$, while observing that

$$P_c^{(N-1)} = (N-1)! \, F_c^{(N-1)}. \tag{2.23}$$

The temporal evolution of the probability density $F^{(N)}$ and, concomitantly, of $P^{(N)}$ is governed by the Liouville equation

$$\partial P^{(N)}(\mathbf{x}^N, t)/\partial t + \mathbf{V}_{\mathbf{x}^N} \cdot [\mathbf{U}^N P^{(N)}(\mathbf{x}^N, t)] = 0, \tag{2.24}$$

expressing the conservation of particles in phase space. Upon integration over $\mathbf{x}_N, \mathbf{x}_{N-1}, \ldots$, similar equations can be derived for each of the lower-order densities $P^{(n)}$, akin to the so-called BBGKY hierarchy (Hansen and McDonald, 1976; McQuarrie, 1976).

The statistical average of any configuration-dependent tensorial quantity $\mathbf{A}(\mathbf{x}^N, t)$ is defined as

$$\bar{\mathbf{A}} = \frac{1}{N!} \int \mathbf{A}(\mathbf{x}^N, t) P^{(N)}(\mathbf{x}^N, t) \, d\mathbf{x}^N. \tag{2.25}$$

Similar conditional averages may also be defined (cf. Section IV).

The incorporation of Brownian motion can be effected in a manner similar to that of Batchelor (1976b), who found the translational diffusion flux due to Brownian motion to be equivalent to one produced by steady forces acting on the particles. In this context, the force exerted on particle i in an ensemble of N particles is taken to be

$$\mathbf{F}_i = -kT \, \partial \ln P^{(N)}(\mathbf{x}^N, t)/\partial \mathbf{x}_i \qquad (1 \leq i \leq N). \tag{2.26}$$

The formal framework introduced in this subsection has been employed, for example, to study the sedimentation of suspensions (Batchelor, 1972).

D. Volume Averages for Homogeneous Suspensions

From an experimental point of view, it is generally more convenient to employ "volumetric" rather than "statistical" [cf. Eq. (2.25)] averages to derive expressions for the bulk properties of suspensions. In order to effect a correspondence between these, an averaging volume V must be selected that is large enough to contain a representative number of particles and yet sufficiently small (compared to the size of the experimental system) such that the local statistical properties of the suspension do not vary appreciably within V. A general multiple-scale analysis will not be introduced here but can be performed as by Brenner (1970) and by Lévy and Sanchez-Palencia (1983a). When such a volume can be found, the ergodicity property of (locally) statistically homogeneous suspensions allows one to equate volumetric and statistical averages.

The volume average of any quantity $\mathbf{A}$ is defined as

$$\langle \mathbf{A} \rangle = V^{-1} \int_V \mathbf{A}\, d^3\mathbf{x}, \tag{2.27}$$

in contrast with Eq. (2.25). Integration of Eq. (2.27) is to be performed over the entire volumetric domain contained within V, including the interior of the particulate phase. It is sometimes useful to replace the latter integral by an equivalent surface integral extended over the external areal domain ∂V bounding V (Brenner, 1970), so as to avoid dealing with possible ambiguities pertaining to the values of $\mathbf{A}$ existing within the particle interiors. When $\mathbf{A}$ is a divergence-free tensor of any order (other than zero), it is easy to show that

$$\langle \mathbf{A} \rangle = V^{-1} \int_{\partial V} \mathbf{x}\, d\mathbf{s} \cdot \mathbf{A}. \tag{2.28}$$

Necessary definitions now exist at this stage to permit continuing the discussion preceding and following Eq. (2.5) for the case of *infinitely* extended suspensions. When N is infinite rather than finite, average macroscopic quantities must be prescribed in place of the prior asymptotic boundary condition (2.5). These must be defined as volume averages. For example, the macroscopic velocity gradient $\langle \mathbf{G} \rangle$ is defined as

$$\langle \mathbf{G} \rangle = V^{-1} \int_V \nabla \mathbf{v}\, d^3\mathbf{x}, \tag{2.29}$$

which is decomposable into respective symmetric $\langle \mathbf{S} \rangle$ and antisymmetric $\langle \ddot{\Lambda} \rangle$ contributions, as in Eq. (2.1).

Among possible macroscopic dynamical quantities of interest, the most useful is the macroscopic stress,

$$\langle \mathbf{P} \rangle = V^{-1} \int_V \mathbf{P}\, d^3\mathbf{x}. \qquad (2.30)$$

As usual, its isotropic portion, corresponding to the average pressure, is first removed as being without consequence for the incompressible flows of interest to us here. Its antisymmetric part, which is related to any external-body couples exerted on the suspended particles, is also separated out. These operations result in the symmetric and traceless average deviatoric stress,

$$\langle \bar{\tau} \rangle = \tfrac{1}{2}(\langle \mathbf{P} \rangle + \langle \mathbf{P}^\dagger \rangle) - \tfrac{1}{3}\mathbf{I}(\mathbf{I}:\langle \mathbf{P} \rangle). \qquad (2.31)$$

To identify the contribution to the latter stress arising from the presence of the particles in the fluid, we write [cf. Eq. (2.8)]

$$\langle \bar{\tau} \rangle = 2\mu_0 \langle \mathbf{S} \rangle + \langle \mathbf{A} \rangle. \qquad (2.32)$$

Elementary calculations (Batchelor, 1970) yield the following expression for $\langle \mathbf{A} \rangle$:

$$\langle \mathbf{A} \rangle = \frac{1}{V} \sum_{i \in V} \ddot{\mathscr{S}}_i, \qquad (2.33\text{a})$$

with

$$\ddot{\mathscr{S}}_i \stackrel{\text{def}}{=} \tfrac{1}{2} \int_{S_i} [d\mathbf{s} \cdot \mathbf{P}\mathbf{x} + \mathbf{x}\, d\mathbf{s} \cdot \mathbf{P} - \tfrac{2}{3}\mathbf{I}(d\mathbf{s} \cdot \mathbf{P} \cdot \mathbf{x})]. \qquad (2.33\text{b})$$

The so-called particle contribution $\langle \mathbf{A} \rangle$ to the mean stress may thus be evaluated by performing an integration over the surfaces of all particles contained within V.

Such a generic derivation was first effected by Landau and Lifshitz (1959) in the absence of any complicating factors. Included in these complicating factors are inertia, which necessitates the introduction of Reynolds stresses, as well as interfacial tension, present when the suspension is composed of droplets rather than rigid particles. Batchelor's (1970) analysis incorporates such factors.

E. Concluding Remarks

The preceding paragraphs furnish the basic elements constituting the starting point for rheological investigations of suspensions. Subsequent analysis based thereon has a two-fold thrust: (i) Determination of the (statistical) generic structure of the suspension and (ii) calculation of the hydrodynamic interaction forces. Of course, these are inseparably intertwined since the forces determine the structure, whereas the structure determines the forces. It is precisely this duality that makes the study of suspensions challenging.

In practice, progress occurs, or at least is deemed to have occurred, when one of these two reciprocal factors has been simplified. In most cases, a priori, ad hoc assumptions are made regarding the suspension's geometric microstructure. In cell models, the configurational problem is artificially eliminated upon replacing the suspension with an "equivalent" collection of space-filling cells, the bounding surfaces of which have kinematical and/or dynamical boundary conditions imposed on them. By their very nature, cell models, however, represent a mean-field treatment of suspension microstructure and, hence, cannot be extended to include interparticle forces, clustering, or other correlation effects. In dilute suspensions, the particulate concentration is assumed to be so low that only single-sphere interaction with the fluid needs to be considered. In statistical models the missing element is the many-body interaction, which is often crudely approximated either as a superposition of two-body interactions or as a power series expansion in inverse powers of the mean-separation distance, with the configuration obtained by a Monte Carlo-type calculation.

Percolation approaches (Section VI) attempt to predict the gross structure of the suspension without, however, requiring detailed knowledge of the interaction forces. Beyond a critical concentration, the existence of an infinite cluster of particles can be expected to occur in shear flows. In spatially periodic models of suspensions (Section VII), one can avoid the problem of structure by assuming the instantaneous particle configuration as well as the kinematical motions of the particles (but not the fluid) to be known a priori, independently of the dynamics. The response of the suspension to the particle motion is studied with the important bonus that the many-body hydrodynamic interaction problem can be solved *exactly* for the spatially periodic configuration. As such, the spatially periodic configuration offers interesting possibilities towards obtaining a complete solution, even when a large number of unequal-size particles is contained within each unit cell of the periodic array.

In Stokesian dynamics (Section VIII), a direct simulation is made starting with a randomly chosen initial particle configuration. The structure is allowed to evolve as part of the detailed fluid-mechanical solution, and the hydrodynamic particle interactions are determined at least to the extent of assuming pairwise additivity of these interactions. The "momentum tracer" method (Section VIII) is characterized by the interesting feature that the particulate phase of the suspension is at rest. The static configuration of this suspension is assumed given, and results are derived for the configuration-specific kinematic viscosity tetradic of the suspension, the latter being viewed as a macroscopic continuum.

III. Empirical and Cell Models

This brief section provides a historical and practical overview of useful empiricisms employed in suspension theories, including a few useful formulas. Early investigators were mainly concerned with the measurement and correlation of two fundamentally important, but apparently unrelated, quantities: (i) The "effective" viscosity μ of sheared suspensions of neutrally buoyant particles and (ii) the sedimentation speed u_s of suspensions of non-neutrally buoyant particles. Upon appropriate normalization, both were regarded as being functions only of the volumetric solids concentration ϕ:

$$\mu/\mu_0 \stackrel{\text{def}}{=} \mu_r(\phi), \tag{3.1a}$$

$$u_s/u_0 \stackrel{\text{def}}{=} u_r(\phi), \tag{3.1b}$$

with μ_r, the so-called relative viscosity, and u_0, the settling velocity of an isolated particle. The forms in Eq. (3.1), where the indicated ratios depend only on ϕ, are now recognized to be correct only in the absence of nonhydrodynamic forces. Subsequent experiments were primarily oriented towards determining the statistical structure of the suspensions (cf. Section V).

Many experimental results on viscosity were empirically correlated by Rutgers (1962a,b) and Thomas (1965). A useful semiempirical formula is that of Mooney (1951). Its derivation is based on an approximate functional argument together with the introduction of an ad hoc "crowding factor" k. Upon contemplating two possible ways of reaching the same total concentration $\phi_1 + \phi_2$, Mooney proposed the formula

$$\mu_r = \exp[\beta\phi(1 - k\phi)^{-1}], \tag{3.2}$$

where β and k are constants. To achieve asymptotic agreement with Einstein's limiting result [see Eq. (4.1)], β is set equal to 2.5. The factor k is usually taken as the reciprocal of the maximum attainable concentration $\phi_{\max}$, namely

$$k = \phi_{\max}^{-1}. \tag{3.3}$$

The infinite viscosity prediction of Eq. (3.2) as ϕ approaches its maximum value is not necessarily correct. In the case of a spatially periodic lattice (Section VII), a rigorous analysis incorporating the time dependence of the relative positions of adjacent spheres in a shear flow provides results counter to the infinite viscosity prediction. [This same conclusion is reached independently by Marrucci and Denn (1985) by heuristic reasoning not limited to spatially periodic model systems.] Furthermore, at least for spatially periodic systems, the actual value of ϕ_{max} depends explicitly on the type of flow under consideration, as discussed in Section VII.

Krieger and Dougherty (1959) proposed the closely-related formula

$$\mu_r = (1 - k\phi)^{-\beta/k}, \qquad (3.4)$$

to which similar criticisms may be addressed.

Recent macroscopic rheological experiments shed light on the microrheology of suspensions. Gadala-Maria and Acrivos (1980) observed the behavior of concentrated suspensions of spheres under several flow conditions. Longtime drift was observed to occur in the value of the shear viscosity, ultimately yielding a reproducible asymptotic value. Under unsteady shear conditions, the macroscopic stress exhibited memory. These phenomena were originally attributed to the development of structure in the suspension for long periods of time; however, Leighton and Acrivos (1987) have recently shown this effect to be due to so-called shear-induced particle diffusion.

Pätzold (1980) compared the viscosities of suspensions of spheres in simple shear and extensional flows and obtained significant differences, which were qualitatively explained by invoking various flow-dependent sphere arrangements. Goto and Kuno (1982) measured the apparent relative viscosities of carefully controlled bidisperse particle mixtures. The larger particles, however, possessed a diameter nearly one-fourth that of the tube through which they flowed, suggesting the inadvertant intrusion of unwanted wall effects.

Similar experiments have been performed with settling suspensions, which leads to comparable empirical correlations. Early experiments are reviewed by Happel and Brenner (1965). Reviews by Fitch (1979) and by Davis and Acrivos (1985) provide the state of the art.

Two formulas widely used to correlate experimental sedimentation data are

$$u_r = (1 - \phi)^\beta \qquad (3.5a)$$

and

$$u_r = \frac{1 - 1.88\phi}{1 + 5\phi}. \qquad (3.5b)$$

The former was devised by Maude and Whitmore (1958) to fit experimental results available at the time, whereas the latter was theoretically derived by Reed and Anderson (1980) upon assuming pairwise additivity of the hydrodynamic particle–particle interactions and by preaveraging such interactions among all particles save the test one. Although not entirely satisfactory from a theoretical viewpoint, Eq. (3.5b) is found to be in good agreement with data (Kops-Werkhoven and Fijnaut, 1982). Buscall et al. (1982) correlated their carefully obtained experimental data by analogy with the Krieger and Dougherty (1959) relative viscosity formula in Eq. (3.4) to obtain

$$u_r = (1 - k\phi)^{-\beta/k}. \qquad (3.5c)$$

Estimates of u_r based on pairwise additivity by Glendinning and Russel (1982) may also be cited. These authors find, however, that the approximate, pairwise-additive treatment of hydrodynamic interactions fails to be adequate for all but very dilute suspensions. Their theoretical formalism can nonetheless be systematically improved by including three- (or more) particle interaction effects.

Cell models constitute a second major class of empirical developments. Among these, only two will be mentioned here as constituting the most successful and widely used. The first, due to Happel (1957, 1958), is useful for estimating the effective viscosity and settling velocity of suspensions. Here, the suspension is envisioned as being composed of fictitious identical cells, each containing a single spherical particle of radius a surrounded by a concentric spherical envelope of fluid. The radius b of the cell is chosen to reproduce the suspension's volume fraction ϕ via application of the formula

$$\phi = (a/b)^3. \qquad (3.6)$$

Stokes equations are solved within the cell, satisfying stick boundary conditions imposed at the particle surface $r = a$ together with vanishing tangential stress on the outer boundary $r = b$ (along with additional kinematical boundary conditions imposed there on the velocity field, serving to isolate the cell from the remainder of the suspension). Theoretical results, containing no adjustable parameters, are in fair agreement with much experimental data.

Cell-type models are still in use (Adler, 1979; Russel and Benzing, 1981) because of their simplicity. Predicted results are often quite reasonable, exhibiting features intuitively anticipated. On the other hand, such models fail to provide definitive answers to many of the fundamental issues encountered in suspension rheology. Moreover, because of their strictly ad hoc geometric nature, no obvious way exists for their rational improvement.

Frankel and Acrivos (1967) introduced a rather different type of cell model based on lubrication-theory-type arguments. First, the rate of mechanical energy dissipation in the small gap between adjacent sphere pairs is calculated, based on two spheres approaching (or receding from) one another along their line of centers. (Dissipation, arising from the relative "sliding motion" occurring between two spheres moving along different streamlines in a simple shear flow, is negligible when compared to that caused by the normal component of the relative motion.) The relative velocity of the two spheres, as well as the time-averaged gap distance between them, is deduced from average values of the shear rate and the particle concentration. All line-of-centers orientations are assumed equally likely, while the average interparticle gap width is inferred from the specified mean-particle concentration (assuming a cubic arrangement of spheres). The relative viscosity thereby obtained (Frankel and Acrivos, 1967) is

$$\mu_r = \frac{9\chi^{1/3}}{8(1 - \chi^{1/3})} \quad \text{as} \quad \chi \stackrel{\text{def}}{=} \frac{\phi}{\phi_{\max}} \to 1. \qquad (3.7)$$

Though Eq. (3.7) compares well with some experimental data, it does not agree with the analysis of Section VII, where it is shown for time- and spatially-periodic models of sheared suspensions that, although the *instantaneous* contribution to the stress tensor is indeed singular, its *time average* is not! Stated alternatively, the contribution arising from the inner "lubrication" zone is of the same order of magnitude as that of the outer nonsingular region. This conclusion can be compared to that of Batchelor and Green (1972b), as commented on by Batchelor (1974), to the effect that the dissipation in the inner zone (arbitrarily defined as a nondimensional gap of less than $0.0025a$ between adjacent sphere surfaces) contributes only about 20% of the effective viscosity to the ϕ^2 term in dilute suspensions. Here again, the contribution arising from the lubrication layer is important but not predominant. Similar remarks relating to Eq. (3.7) have been made by Marrucci and Denn (1985). Further comments on Eq. (3.7) are given in Section VIII,A, where the two-dimensional counterpart of Eq. (3.7) is found to agree with dynamic simulations by Brady and Bossis (1985), although for reasons unrelated to those underlying the Frankel and Acrivos (1967) derivation.

Indirectly related to the cell models of this section is the work of Davis and Brenner (1981) on the rheological and shear stability properties of *three-phase* systems, which consist of an emulsion formed from two immiscible liquid phases (one, a discrete phase wholly dispersed in the other continuous phase) together with a third, solid, particulate phase dispersed within the interior of the discontinuous liquid phase. An elementary analysis of droplet breakup modes that arise during the shear of such three-phase systems reveals that the destabilizing presence of the solid particles may allow the technological production of smaller size emulsion droplets than could otherwise be produced (at the same shear rate).

IV. Dilute Suspensions

As is well known, Einstein (1906, 1911) calculated the additional rate of mechanical energy dissipation engendered by the introduction of a single sphere into a homogeneous shear flow and ultimately obtained

$$\mu/\mu_0 = 1 + \tfrac{5}{2}\phi \qquad (\phi \ll 1) \tag{4.1}$$

for the effective viscosity μ of a dilute suspension. (Interestingly, the existence of two dates associated with this result stems from an algebraic slip committed by Einstein in his original 1906 paper and subsequently corrected in 1911.)

Later, Landau and Lifshitz (1959) obtained the same result by averaging the stress tensor over the entire space, thereby initiating one of the first dynamical (i.e., nonenergetic) approaches to calculating the rheological properties of suspensions. Attempts to extend Eq. (4.1) to higher concentrations are legion. Most propose a power series expansion of the form

$$\mu/\mu_0 = 1 + 2.5\phi + k_1\phi^2 + k_2\phi^3, \ldots. \tag{4.2}$$

The long-range, purely hydrodynamic interaction between two suspended spheres in a shear flow was first calculated by Guth and Simha (1936), yielding a value of $k_1 = 14.1$ via a reflection method. Saito (1950, 1952) proposed two alternative modifications, obtaining $k_1 = 12.6$ and 2.5, respectively; the latter value is obtained upon supposing a spatially uniform distribution of particles.

Vand's (1948) analysis considered the effect of adding an incremental volume fraction $\delta\phi$ of spheres to an existing suspension of spheres of concentration ϕ. Also included was the contribution arising from doublets. His value of $k_1 = 7.35$ was later corrected by Manley and Mason (1954) to $k_1 = 10.05$. In a pioneering, often overlooked, investigation, Peterson and Fixman (1963) obtained $k_1 = 4.31784$, correctly identifying and properly posing many of the underlying questions later addressed and resolved more rigorously by others. Their approximate result was obtained by assuming a uniform density for the two-particle probability density and using an inaccurate far-field approximation for the hydrodynamic interaction.

In a systematic study of particle-particle interactions that also has often been overlooked, Cox and Brenner (1971) developed a comprehensive general theory for calculating the rheological properties of a suspension of particles of arbitrary shape to $0(\phi^2)$, including wall effects. They did not, however, attempt an explicit numerical calculation of k_1, and it is perhaps for this reason that their work has not received the thoughtful attention it deserves.

Finally, Batchelor and Green (1972b) furnished a rigorous calculation of the coefficient k_1. In a subsequent series of papers, Batchelor also examined the sedimentation of a dilute suspension (Batchelor, 1972), the effects of Brownian motion (Batchelor, 1976b, 1977), and polydispersity effects (Batchelor, 1982, 1983; Batchelor and Wen, 1982). In what follows, Batchelor and Green's (1972b) derivation of the bulk stress to order ϕ^2 will be examined, emphasizing the key point of their arguments. Subsequently, other relevant results, as well as other derivations, will be quoted and extensions thereof will be cited.

Consider a suspension of identical, spherical (radii a), force- and couple-free particles. Upon neglect of both inertia and Brownian movement, the proper rheological starting point is Eq. (2.33) for the particle stress. As in Eq. (4.1), the contribution arising when particle–particle interactions are absent is known. Its explicit inclusion in $\langle \mathbf{A} \rangle$ yields

$$\langle \mathbf{A} \rangle = 5\phi\mu_0 \langle \mathbf{S} \rangle + 5\phi\mu_0 \left(\frac{\langle \mathbf{A} \rangle}{5\phi\mu_0} - \langle \mathbf{S} \rangle \right). \tag{4.3}$$

The second term reflects interaction effects. The parenthetical contribution appearing in the latter term may be alternatively expressed as

$$\frac{\ddot{\mathscr{S}}}{20\pi a^3 \mu_0 / 3} - \langle \mathbf{S} \rangle, \tag{4.4}$$

with $\ddot{\mathscr{S}}$ as the statistical average of $\ddot{\mathscr{S}}_i$ [cf. Eq. (2.33b)] for the averaging volume V containing N particles. As such, Eq. (4.4) is equivalent to

$$\frac{1}{N!} \int \left(\frac{\ddot{\mathscr{S}}}{20\pi a^3 \mu_0 / 3} - \langle \mathbf{S} \rangle \right) P^{(N)}(\mathbf{x}^N) d\mathbf{x}^N, \tag{4.5a}$$

which may be transformed to the form

$$\frac{1}{(N-1)!} \int \left(\frac{\ddot{\mathscr{S}}}{20\pi a^3 \mu_0 / 3} - \langle \mathbf{S} \rangle \right) P_c^{(N-1)}(\mathbf{x}^{N-1} | \mathbf{x}_0) d\mathbf{x}^{N-1} \tag{4.5b}$$

with the aid of Eq. (2.2). Here, $\ddot{\mathscr{S}}$ is to be regarded as the particle stress, functionally dependent on the geometric arrangement of the $N-1$ remaining particles, given that the designated particle is located at $\mathbf{x}_0$.

A number of difficulties arise in attempting to exploit Eq. (4.5b). First, the conditional density $P_c^{(N-1)}$ is as yet unknown, apart from being difficult to obtain. This may be circumvented by supposing that interaction with one particle constitutes the dominant contribution to Eq. (4.5b), allowing the averaging to be performed by replacing the weighting function,

$$[(N-1)!]^{-1} P_c^{(N-1)} d\mathbf{x}^{N-1},$$

in Eq. (4.5b) by the simpler, two-particle conditional probability, $P(\mathbf{x}_0 + \mathbf{R} | \mathbf{x}_0) d^3 \mathbf{R}$. However, this probability tends to the position-independent value N/V for large $|\mathbf{R}|$. The concomitant $0(|\mathbf{R}|^{-3})$ integrand results in a non-absolutely convergent integral. To deal with this difficulty, Batchelor and Green (1972b) proposed finding a quantity whose mean value is known and which for large $|\mathbf{R}|$ possesses the same functional dependence as does the integrand. By its subtraction from and addition to the integrand, proper evaluation of the integral can be effected.

The required conditional probability density $P(\mathbf{x}_0 + \mathbf{R} | \mathbf{x}_0)$ for finding a second particle at $\mathbf{x}_0 + \mathbf{R}$ (given the existence of the reference sphere at $\mathbf{x}_0$) can be obtained by solving the Liouville equation,

$$\frac{\partial}{\partial t} P(\mathbf{x}_0 + \mathbf{R} | \mathbf{x}_0) + \frac{\partial}{\partial \mathbf{R}} \cdot [\mathbf{V}(\mathbf{R}) P(\mathbf{x}_0 + \mathbf{R} | \mathbf{x}_0)] = 0 \qquad (4.6)$$

[with $\mathbf{V}(\mathbf{R})$ the relative velocity of the centers of two spheres], subject to the boundary conditions

$$P = 0 \quad \text{at } R = 2a \qquad (4.7a)$$

and

$$P \to N/V \equiv 3\phi/4\pi a^3 \quad \text{as } R \to \infty, \qquad (4.7b)$$

applicable for systems without long-range order. Despite their widespread use, both of the above boundary conditions are of a tentative hypothetical nature; different results would, of course, arise from the choice of other conditions (Yoshida, 1988). Furthermore, Eq. (4.6) shows the probability to be simply convected with the velocity $\mathbf{V}$. It is therefore possible that some domains of the flow are not reached by the streamlines coming from infinity. This phenomenon occurs, for instance, in simple shear flow (Darabaner and Mason, 1967). Hence, in such circumstances, only particular probability densities are obtained, without any universal meaning.

To obtain $\mathscr{S}$, it is further necessary to solve the Stokes problem for two spheres. In the case of a purely straining motion, where only open trajectories obtain the result of the complete analysis, as outlined above, eventually yields (Batchelor and Green, 1972b)

$$\mu/\mu_0 = 1 + \tfrac{5}{2}\phi + 7.6\phi^2. \qquad (4.8)$$

This can be compared to the previously cited results pertaining to Eq. (4.2). Yoon and Kim (1987) have more accurately recalculated the last coefficient in Eq. (4.8) to be 6.95.

Similar methods have been applied by Batchelor (1972, 1982; Batchelor and Wen, 1982) to a number of other problems. A sedimenting suspension of equal spheres was examined first. Since no relative motion exists between two identical settling spheres, the probability density cannot be determined from

Eq. (4.6); rather, the distribution was assumed uniform, and the mean sedimentation velocity u_s of a settling sphere found to be related to the velocity u_0 of an isolated sphere by the expression

$$u_s/u_0 \simeq 1 - 6.55\phi. \tag{4.9}$$

This result has been generalized (Batchelor, 1982; Batchelor and Wen, 1982) to a polydisperse suspension of settling spheres, obtaining for the average settling velocity $u_{s,i}$ of species i,

$$u_{s,i}/u_{0,i} = 1 + \sum_{j=1}^{m} s_{ij}\phi_j, \tag{4.10}$$

with m, the total number of species; $u_{0,i}$, the settling speed of a particle of species i in isolation; and ϕ_j, the volume fraction of species j. Calculated values of the coefficients s_{ij} are displayed in the previous reference for a variety of circumstances. For spherical particles, there is no influence manifested of translational Brownian motion on the first-order Einstein coefficient in Eq. (4.1)*. Such effects do, however, arise at $O(\phi^2)$ as a consequence of the particle–particle interactions.

In a sheared suspension, the effects are two-fold. First, the expression for bulk stress itself must be modified. Second, the probability density is affected since the continuity equation for the latter must be replaced by a convection–diffusion equation. As a consequence, the distinction between open and closed trajectories loses some of its meaning. Batchelor (1977) gives the equivalent viscosity of a sheared suspension subject to strong Brownian motion as

$$\mu/\mu_0 = 1 + \tfrac{5}{2}\phi + 6.2\phi^2, \tag{4.11}$$

independently of the explicit form adopted for the bulk flow. The current status of the contribution of Brownian motion to the viscosity of suspensions of spherical particles is reviewed by Felderhof (1989).

Finally, the mean flux of particles of various sizes in the presence of bulk concentration gradients was calculated (Batchelor, 1976b, 1983), and the corresponding Fickian diffusion tensors systematically obtained. With no interparticle forces, the numerical results may be represented by the approx-

---

* However, for nonspherical particles, rotational Brownian motion effects already arise at $O(\phi)$. In the case of ellipsoidal particles, such calculations have a long history, dating back to early polymer-solution rheologists such as Simha and Kirkwood. Some of the history of early incorrect attempts to include such rotary Brownian effects is documented by Haber and Brenner (1984) in a paper addressed to calculating the $O(\phi)$ coefficient and normal stress coefficients for general triaxial ellipsoidal particles in the case where the rotary Brownian motion is dominant over the shear (small rotary Peclet numbers)—a problem first resolved by Rallison (1978).

imate interpolation formulas

$$D_{ii}/D_i^{(0)} = 1 + 1.45\phi_i - \sum_{k(\neq i)} \frac{2.5\phi_k}{1 + 0.6\lambda_{ik}} \quad (4.12a)$$

$$D_{ij}/D_i^{(0)} = \phi_i(\lambda_{ij}^3 + 2\lambda_{ij}^2) \quad (j \neq i), \quad (4.12b)$$

in which $\lambda_{ij} = a_j/a_i$. The effect of colloidal forces was also explored in the cited references, and numerical results were obtained for a few representative cases.

Transcending the central body of knowledge for dilute suspensions described in the preceding paragraphs, Russel (1976, 1978) analyzed the rheological properties of electrostatically stabilized colloidal suspensions, including Brownian motion. Details are provided in limiting cases. Haber and Hetsroni (1981) investigated the sedimentation of liquid spheres of different sizes, whose density and viscosity differed from that of the suspending fluid. Feuillebois (1984) derived the average sedimentation velocity of monodisperse solid spheres in the presence of vertical concentration inhomogeneities. Caflish and Luke (1985) calculated the variance of the sedimentation speed; in the infinite particle number limit, it was found to be infinite, a conclusion whose physical significance is unclear.

Comparisons with experimental data have been effected by Batchelor. In careful experiments, Kops-Werkhoven and Fijnaut (1981) obtained -6 ± 1 for the coefficient of the first-order volume fraction term in the sedimentation velocity [cf. Eq. (4.9)] and 1.3 ± 0.2 for the corresponding term in the diffusion coefficient [cf. Eq. (4.12)]. The latter term is derived from light-scattering experiments. For monodisperse suspensions, these results agree reasonably well with Batchelor's predictions.

In related contributions, Felderhof (1978) arrived at this same diffusion coefficient, except for a numerical difference due to his use of a different far-field expression for the hydrodynamic interaction. Several investigators proposed schemes for circumventing the difficulty that arises from the existence of nonabsolutely convergent integrals in the theory. Thus, Jeffrey (1974) employed a group expansion method; Hinch (1977) statistically averaged the conservation and constitutive relations, obtaining an infinite hierarchy of equations which then had to be truncated; O'Brien (1979) used an integral representation of the solution to the basic equations; Feuillebois (1984) devised a novel method based on decoupling the treatment of the divergent integrals from the calculation of hydrodynamic interactions, while keeping the reference volume finite.

Another related field of interest—but one beyond the scope of our review—is the concentration dependence of the self and collective diffusivities, including distinctions between short- and long-term self diffusivities. The interested reader is referred to the careful discussions by Batchelor (1983), Rallison and Hinch (1986) and Yoshida (1988) in this context.

V. Statistical Models

Theoretical trends in the study of suspensions employ concepts and techniques originally developed in connection with theories of liquids, for example, equation hierarchies, closure problems, and Monte Carlo methods. In marked contrast with the definitive achievements reviewed in the previous section, the present section outlines a field currently under active development.

Knowledge of basic molecular–theoretic (Hansen and McDonald, 1976) concepts and properties, such as those of hard-sphere models of liquids, constitutes a prerequisite for admission to the present field. While relevant experimental techniques in this field will not be reviewed here, it is difficult to avoid mentioning light scattering. The review by Pusey and Tough (1982) shows how dynamic light scattering can provide data on both collective and self-diffusion coefficients. Moreover, the static structure factor, which is the spatial Fourier transform of the radial distribution function, can be obtained by static light-scattering experiments. Discovery of many enlightening results relating to suspension structures can be anticipated with further development of this experimental technique. A paper by Rallison and Hinch (1986) provides an illuminating discussion of dynamic light-scattering phenomena, with emphasis on particle-interaction effects.

Various formal expansions are outlined below, followed by a summary of numerical calculations employing Monte Carlo techniques and major conclusions derived from these techniques.

A. Formal Expansions

The review by Herczyński and Pieńkowska (1980) of these expansions provides an excellent entrée to the major theories analyzed in the literature on the subject, most of which address determination of the permeability of random arrays of spheres. In this context, it is useful to define the so-called phase function,

$$H(\mathbf{r}, t) = \begin{cases} 1 & \text{in the fluid,} \\ 0 & \text{in the particles,} \end{cases} \tag{5.1}$$

which may be used, for example, to calculate averages restricted to the fluid phase alone.

As before, let **P** be the local stress tensor, and denote by an overbar the statistical average of any quantity. The definition of the fluid-velocity field may be analytically extended to the solid-particle interiors and the pressure therein assumed to vanish. As such, taking the statistical average of the

Stokes equations and using the phase function $H(\mathbf{r}, t)$, it may be shown formally that

$$-\nabla \overline{p(\mathbf{r})} + \mu_0 \nabla^2 \overline{\mathbf{v}(\mathbf{r})} = n(\mathbf{r}) \int_{s_1} d\mathbf{s} \cdot \overline{\mathbf{P}(\mathbf{r} | \mathbf{R}_1)}, \qquad (5.2a)$$

with $n(\mathbf{r}, t)$, the particle-number density function; $\mathbf{P}(\mathbf{r} | \mathbf{R}_1)$, the stress tensor existing when a sphere is centered at $\mathbf{R}_1$; and s_1, the surface of the latter sphere. Complementing this equation is the averaged continuity equation,

$$\nabla \cdot \overline{\mathbf{v}(\mathbf{r})} = 0. \qquad (5.2b)$$

Appearing in the integrand of Eq. (5.2a) is the averaged stress,

$$\overline{\mathbf{P}(\mathbf{r} | \mathbf{R}_1)} = \overline{p(\mathbf{r} | \mathbf{R}_1)} \mathbf{I} + \mu_0 [\overline{\nabla \mathbf{v}(\mathbf{r} | \mathbf{R}_1)} + \overline{\nabla \mathbf{v}^\dagger(\mathbf{r} | \mathbf{R}_1)}]. \qquad (5.3)$$

The required equations for $\overline{p(\mathbf{r} | \mathbf{R}_1)}$ and $\overline{\mathbf{v}(\mathbf{r} | \mathbf{R}_1)}$ are derived by conditionally averaging the Stokes equations, thereby obtaining

$$-\nabla \overline{p(\mathbf{r} | \mathbf{R}_1)} + \mu_0 \nabla^2 \overline{\mathbf{v}(\mathbf{r} | \mathbf{R}_1)} = n(\mathbf{r} | \mathbf{R}_1) \int_{s_2} d\mathbf{s} \cdot \overline{\mathbf{P}(\mathbf{r} | \mathbf{R}_1, \mathbf{R}_2)}, \qquad (5.4a)$$

$$\nabla \cdot \overline{\mathbf{v}(\mathbf{r} | \mathbf{R}_1)} = 0. \qquad (5.4b)$$

Equations (5.4a,b) are similar to Eqs. (5.2) except for the additional conditioning with respect to $\mathbf{R}_1$. To evaluate the RHS one must obtain equations for $P(\mathbf{r} | \mathbf{R}_1, \mathbf{R}_2)$ and $\mathbf{v}(\mathbf{r} | \mathbf{R}_1, \mathbf{R}_2)$ by twice conditionally averaging the Stokes equations.

The hierarchy of equations thereby obtained can be closed by truncating the system at some arbitrary level of approximation. The results eventually obtained by various authors depend on the implicit or explicit hypotheses made in effecting this closure—a clearly unsatisfactory state of affairs. Most contributions in this context aim at calculating the permeability (or, equivalently, the drag) of a porous medium composed of a random array of spheres. The earliest contribution here is due to Brinkman (1947), who empirically added a Darcy term to the Stokes equation in an attempt to represent the hydrodynamic effects of the porous medium. The so-called Brinkman equation thereby obtained was used to calculate the drag exerted on one sphere of the array, as if it were embedded in the porous medium continuum. Tam (1969) considered the same problem, treating the particles as point forces; he further assumed, in essence, that the RHS of Eq. (5.2a) was proportional to the average velocity and hence was of the explicit form

$$-\alpha^2 \mu_0 \overline{\mathbf{v}(\mathbf{r})}. \qquad (5.5)$$

The phenomenological constant α is calculated by Brinkman's self-consistent argument. In circumstances where the particles are identical, the drag computed in this manner reduces to Brinkman's formula.

Lundgren (1972) treated the case of a suspension of spheres. He assumed the RHS of Eq. (5.2a) to be a linear functional of $\overline{\mathbf{v}(\mathbf{r})}$, choosing this functional dependence to be of the form

$$A\overline{\mathbf{v}(\mathbf{r})} + B\nabla^2\overline{\mathbf{v}(\mathbf{r})}. \tag{5.6}$$

The first term is, of course, equivalent to that considered by the previous authors, whereas the second combines with the Stokes term to yield an effective viscosity. Coefficients A and B were determined through a self-consistent scheme.

Howells (1974) restricted his attention to fixed particles, extending the method of Childress (1972) by considering a given number of particles chosen from an infinite set. This partly self-consistent scheme furnishes terms valid in the small-solids concentration limit. In a very readable paper, Hinch (1977) combined some of the above procedures in formulating an "averaged-equation" approach to particle interactions, providing expressions for the bulk stress, average sedimentation velocity, and effective permeability in suspensions and fixed beds.

In the same vein, Itoh (1983) employed Eq. (5.5) to calculate α using the same self-consistent argument, but with the inclusion of an intermediate fluid layer around the test sphere. Theoretical results compared well with experimental results.

Another series of related papers includes those of Herczyński and Pieńkowska (1980), who reinterpreted the derivation of Bedeaux et al. (1977) in terms of grand-canonical averages. Buyevich and coworkers developed yet another scheme by spatially averaging the equations in each phase of the suspension; a complete account of this can be found in Buyevich and Shchelchkova (1978). Contributions by Freed and Muthukumar provide yet another perspective. As their first entrée (Freed and Muthukumar, 1978), they considered the Stokes problem for the velocity disturbance and friction coefficient arising from the motion of a single sphere moving with constant velocity U through a stationary suspension of spheres at finite concentration. They obtained a formal solution given in terms of the so-called Oseen tensor. These authors were not concerned with the velocity field per se, but rather only with its statistical average. Divergent integrals were encountered and summed, and the average velocity field was shown to be a solution of the Brinkman equation. A number of applications exist for these results; however, in view of the formidable nature of the expansion, calculations are generally restricted to second-order terms, corresponding to binary particle–particle interactions.

Freed and Muthukumar (1982) and Muthukumar and Freed (1982) applied a similar scheme to suspensions of hard spheres. Results resembling those of Batchelor and Green (1972b) were obtained upon assuming a spatially uniform distribution of spheres.

Beenakker (1984) and Beenakker and Mazur (1983, 1984) used their method of induced forces (cf. Section II) to calculate particle diffusivity and effective viscosity in concentrated suspensions. Beenakker's (1984) paper introduces a novel wave vector-dependent viscosity $\mu(\mathbf{k})$ by considering the average response of the suspension to an externally applied force. The usual viscosity is obtained by considering the limit of zero wave-number, from which an explicit expression for μ (as a double summation of propagators) is obtained. This general formula was used in various ways. For dilute suspensions, the viscosity was expanded in terms of the concentration, yielding the same results as those of Freed and Muthukumar (cf. the previous paragraph). Another original feature entails the procedure used at high concentrations: $\mu(k)$ is expressed in terms of renormalized connectors in order to account for spheres interacting with one another via the intervening suspension. The effective viscosity is then expanded with the help of density-fluctuation correlation functions of increasing order, wherein the first two terms of the expansion are evaluated via the equilibrium pair distribution function $g(r)$ (i.e., the classical Percus-Yevick distribution for hard spheres). Agreement between numerical and experimental results is satisfactory for solids concentrations of up to 30% (see de Kruif et al., 1985). It must be stressed, however, that the above treatments account only for the viscous contribution to the stress, while neglecting the effect on the bulk stress tensor of the nonequilibrium structures formed in sheared suspensions.

B. Numerical Calculations

Monte Carlo techniques were first applied to colloidal dispersions by van Megen and Snook (1975). Included in their analysis was Brownian motion as well as van der Waals and double-layer forces, although hydrodynamic interactions were not incorporated in this first study. Order disorder transitions, arising from the existence of these forces, were calculated. Approximate methods, such as first-order perturbation theory for the disordered state and the so-called cell model for the ordered state, were used to calculate the latter transition, exhibiting relatively good agreement with the "exact" Monte Carlo computations. Other quantities of interest, such as the radial distribution function and the excess pressure, were also calculated. This type of approach appears attractive for future studies of suspension properties.

van Megan *et al.*, (1983) and Pusey and van Megan (1983) employed the expansion proposed by Mazur and van Saarloos (1982) to obtain diffusion coefficients in concentrated, hard-sphere dispersions. Comparison of these predictions with experimental data for relatively concentrated suspensions revealed that two- and three-particle interactions were of the same orders of magnitude. Inclusion of two-body terms, however, made the agreement with experiments worse, whereas the further addition of three-body terms improved the comparison. This result is intriguing since Beenakker and Mazur (1982) have recently shown that three-body contributions change the algebraic sign of the second-order term in the expression for the concentration-dependence of the self-diffusion coefficient. They conclude that "... two-sphere hydrodynamic interactions do not suffice to describe the properties of suspensions at higher densities." Of course, this statement also casts a shadow on the validity of analyses where pairwise additivity is assumed.

Other review articles relating to colloidal suspensions and containing discussions of numerical calculations (Monte Carlo simulations) include contributions by van Megen and Snook (1984) and Castillo *et al.* (1984). The scope of these articles is not, however, limited to numerical techniques; they also provide general reviews of the statistical mechanics of colloidal suspensions.

VI. Percolation Models

Application of percolation theory concepts to the study of suspensions has attracted much interest. Major aspects of the general theory are briefly recounted in the following subsection, after which, an analogy between percolation and suspension flows is detailed together with its main predictions. We conclude with a discussion of supporting experimental results obtained for "two-dimensional" suspensions.

A. Percolation Theory

Percolation theory is reviewed by Stauffer (1985). The so-called "site" percolation problem may be posed as in the pioneering analysis by Broadbent and Hammersley (1957), who initiated the subject. If "atoms" are distributed randomly at the sites of a regular lattice in such a way that any given site has probability p of being occupied, what is the probability $P(p)$ that a given atom belongs to an infinite cluster? By an infinite cluster, a connected collection of atom-containing neighboring sites extending to infinity without interruption is meant. The "bond" percolation problem is rather similar to the previous one. Consider a regular lattice and let p now denote the fraction of bonds between two sites which are "favorable" (e.g., not blocked, open, etc., depending on the particular problem being studied). What is the probability $P(p)$ that a favorable bond is part of an infinite cluster linked by such bonds?

Both problems share the common property that $P(p)$ is of measure zero for $p < p_c$, with the critical threshold value p_c as a function of the type of lattice considered. Few exact formulas exist for $P(p)$ or even p_c. There are, however, a number of empirical rules. For instance, for the bond percolation problem,

$$zp_c \approx \frac{d}{d-1}, \tag{6.1}$$

where z is the coordination number and d is the dimensionality of the space in which the lattice is embedded. Near the critical concentration p_c, the probability $P(p)$ adopts the power-law form

$$P(p) = \text{const}|p - p_c|^\beta. \tag{6.2}$$

In two dimensions (d = 2), the so-called "critical exponent" is $\beta \approx 0.14$. Considerably more details (cf. Stauffer, 1985; Havlin and Ben-Arraham, 1983) are known about the geometry of random networks.

Conductance of such a random network constitutes the simplest transport problem that can be studied by percolation ideas. A crude effective-medium theory shows that the bulk conductance is proportional to the number of active bonds:

$$\sigma(p) = \text{const}(p - p_c)^{+1} \quad (p > p_c). \tag{6.3}$$

This expression is confirmed by computer simulations except near the critical value, where it is found that

$$\sigma(p) = \text{const}(p - p_c)^s \quad (p \gtrsim p_c). \tag{6.4}$$

Values ascribed to the critical exponent s in the literature are somewhat controversial. Results (Derrida and Vannimenus, 1982) indicate that s is close to 1.28. Below p_c the bulk conductance is zero, since the network fails to form an infinite cluster whose existence would permit current to be conducted through the system.

According to Stauffer (1979), "A complete understanding of percolation would require [one] to calculate these exponents exactly and rigorously. This aim has not yet been accomplished, even in general for other phase transitions. The aim of a scaling theory as reviewed here is more modest than complete understanding: We want merely to derive relations between critical exponents." Three principal methods currently employed to derive critical exponents are (i) series expansions, (ii) Monte Carlo simulation, and

(iii) renormalization-group theory. Stauffer (1979) outlines the scope, possibilities, and limitations of each. Many (computer generated) results in this area are either empirical or semiempirical; some are obtained under rather delicate conditions. Consequently, knowledge of the existence of past failures, as well as successes, plays an important role in advancing the subject.

B. Application to Suspensions

DeGennes (1979, 1981) was the first to apply percolation concepts to suspensions during an examination of the mechanisms of "collision" occurring between two neutrally buoyant spheres in a simple shear flow—a subject earlier investigated exhaustively by Mason and co-workers (Arp and Mason, 1977b). Compared to molecular collisions occurring in gases, the distinguishing feature of such shear-induced collisions is its finite duration (the time interval being inversely proportional to the shear rate γ). Of course, this gross attribute does not differentiate between the significant differences distinguishing equatorial and polar collision times. Furthermore, from a hydrodynamic viewpoint, no real physical contact occurs between the spheres, whence the precise meaning of the term "collision" is elusive. Nevertheless, the average collision time is reckoned to be of the order of $2.5\gamma^{-1}$. As such, a finite probability p exists for a given sphere to be in "contact" with one (or more) of the others. This probability is independent of γ and is obviously a monotone increasing function of the solid-phase concentration ϕ. Thus, as ϕ and p increase, each sphere becomes progressively more likely to exist as part of a cluster than in isolation.

By analogy to percolation arguments, a critical concentration ϕ_c is assumed to exist, beyond which value an infinite cluster forms. With ϕ_∞ the fraction of spheres belonging to an infinite cluster, the percolation analogy with Eq. (6.2) suggests that

$$\phi_\infty = \text{const}(\phi - \phi_c)^{\beta_3}. \tag{6.5}$$

Formation of such an infinite cluster appears likely to change the suspension hydrodynamics drastically. For example, the classical parabolic Poiseuille velocity profile existing within a circular tube is likely to be severely blunted, a phenomenon observed experimentally by Karnis *et al.* (1966).

In parallel with conductive transport [cf. Eq. (6.4)], momentum transport, as quantified by the shear viscosity, is expected to obey the scaling law

$$\mu/\mu_0 = \text{const}|\phi - \phi_c|^{-s_3}. \tag{6.6}$$

As emphasized by DeGennes (1979), none of the quantitative "predictions" of this percolation model agree with known experimental results other than the apparent formation of an infinite cluster. For instance, both the detailed velocity profile and concentration dependence of the suspension viscosity are at odds with the above predictions, as evidenced by comparison with the early experiments of Karnis et al. (1966).

DeGennes' suggestions were later implemented by the Marseille group. Their experiments were aimed at verifying both the formation of an infinite cluster and the scaling law in Eq. (6.6). The most representative of these contributions is embodied in the work of Bouillot et al. (1982). Ease of observation was achieved by shearing a "two-dimensional" suspension in a macro-couette apparatus. This configuration was achieved by pouring a thin layer of light liquid onto a denser immiscible liquid, following which, equal-sized spheres were inserted into that layer. Admittedly, such experiments suffer a number of drawbacks, including "end effects" arising from the finite thickness of the upper layer and an insufficiency of sphere numbers to assure statistical reliability. Nevertheless, interesting features appear to be present in the experiments.

Here again, as in the three-dimensional "touching" configuration referred to previously, no remarkable "kink" or discontinuity was observed in the viscosity–concentration curve; the results could be correlated by an empirical Eilers (1941) -like law. However, statistical evaluation of the clusters proved interesting. Clusters were defined arbitrarily by supposing that any two spheres closer than $\frac{1}{10}$ of their common radius (corresponding to the resolution of the photographic device) belong to the same cluster. As a function of the "two-dimensional" surface concentration ϕ^s (defined as the areal fraction of the projected area of the spheres to the total area of the Couette apparatus), experiments revealed the appearance of a cluster at about $\phi_c^s \approx 0.65$. The latter embodied a large fraction ($\approx 75\%$) of the total sphere population. The cluster completely surrounded the inner cylinder. Thus, the existence of infinite clusters appears to have been experimentally verified, although other statistical characterizations, such as measurement of the radial distribution function $g(r)$, have yet to be performed.

The foregoing results may be discussed in terms of spatially periodic suspensions, which represent the only exactly analyzable suspension models currently available for concentrated systems. Since spatially periodic models are discussed in the next section, the remainder of this section may be omitted at first reading.

First, the observed critical concentration $\phi_c^s \approx 0.65$ may be compared with the maximum kinematic concentration ($\phi_{max} = \pi/4 \simeq 0.785$) possible for a two-dimensional suspension of circular disks undergoing simple shear. That the actual ϕ_c^s value is lower than the theoretically predicted one may be rationalized in terms of spatially periodic packings allowing the existence of more concentrated systems than disordered packings. According to Berryman

(1983), random close packing (rcp) is predicted to occur at a concentration $\phi^{\text{rcp}} \approx 0.82$ in two dimensions. The theoretical value corresponding to ordered close packing (ocp) in the static case is $\phi^{\text{ocp}} = 0.9069$. The assumption that the randomness causes the same fractional decrease to occur in the dynamic packing case leads to the estimate

$$\phi_c^s = (\phi^{\text{rcp}}/\phi^{\text{ocp}})\phi_{\text{max}} \approx 0.71, \qquad (6.7)$$

which lies within 10% of the observed 0.65 value. A further distinction between random-loose and random-close packing may yet improve the estimate further.

Second, the spatially periodic model suggests further interpretations and experiments. That no "kink" exists in the viscosity vs. concentration curve may be related to the fact that the average dissipation rate remains finite at the maximum kinematic concentration limit, ϕ_{max}. Infinite strings of particles are formed at this limit. It may thus be said that although the geometry "percolates," the resulting fields themselves do not, at least not within the context of the spatially periodic suspension model.

Systematic four-roller (Bentley and Leal, 1986a,b) experiments suggest themselves as candidates to investigate more completely the entire class of two-dimensional shear flows, these being distinguished by a single scalar parameter λ [cf. Eq. (7.8) and Fig. 1]. Measurement of ϕ_c^s as a function of λ, and its subsequent comparison with the predicted $\phi_{\text{max}}(\lambda)$ values, would prove equally interesting. It may also be instructive to exceed $\phi_{\text{max}}(\lambda)$ by changing the flow parameter λ at a fixed concentration ϕ. This would permit the singular case of simple shear flow to be embedded into the more general class of two-dimensional incompressible flows.

An experiment by Camoin and Blanc (1985), performed in the same apparatus, allowed the structure of the resulting clusters to be analyzed in the presence of attractive forces existing between the suspended particles. The resulting structure was found to be fractal in nature for circumstances wherein the attractive forces dominated over hydrodynamics forces.

VII. Spatially Periodic Suspensions

Spatially periodic models of suspensions (Adler and Brenner, 1985a,b; Adler et al., 1985; Zuzovsky et al., 1983; Adler, 1984; Nunan and Keller, 1984) constitute an attractive subject for theoretical treatment since their geometrical simplicity permits rigorous analysis, even in highly concentrated systems. In particular, when a unit cell of the spatially periodic arrangement contains but a single particle, the underlying kinematical problems can be readily identified and solved. Dynamical properties, such as interstitial velocity, pressure, and stress fields, can be derived from these kinematical results using classical tools such as the grand resistance matrix (Brenner and O'Neill, 1972). Formally, the single particle ("perfect crystal") problem has been completely solved, as seen shortly. An immediate application of the single particle problem is to the study of colloidal crystals, which are thought to mimic the microscopic structures and phases observed in atomic crystalline systems. The existence of various length scales provides for easy experimental observation of some spectacular macroscopic phenomena. Overbeek (1982) provides a general view of the physicochemical and colloidal aspects of monodisperse suspensions; Pieranski (1983) emphasizes the solid-state physics of such "crystals." Also of interest in this context is the treatise edited by Pieranski and Rothen (1985).

When more than one particle is contained within a unit cell, the fundamental conceptual problems are essentially the same, though the detailed analyses are considerably more difficult because of the intrusion of nontrivial dynamical features into what were previously purely kinematical ones. Nonetheless, some of the interstitial fluid field-properties, as well as the existence of a grand resistance matrix formulation governing the configurational evolution of the particulate phase, continue to remain valid. The scope of applications is, moreover, greatly enhanced. Some motions of so-called colloidal crystals can only be addressed by assigning more than one particle per cell. Moreover, such multi-particle analyses provide the basis for further suspension-simulation studies via methods similar to those employed in molecular dynamics. In particular, these involve temporal evolutionary studies of the instantaneous spatial distribution of N particles moving within a unit cell and are subject to periodic boundary conditions [see Hansen and McDonald (1976) as well as Section VIII of this chapter]. For the most part this section focuses on our own work on this subject (Adler and Brenner, 1985a,b; Adler et al., 1985; Zuzovsky et al., 1983; Adler, 1984). An independent approach, substantially different in scope, has been developed using so-called homogenization methods (Bensoussan et al., 1978; Sanchez-Palencia, 1980), whose applications have been directed primarily towards immobile particulate systems (such as porous media). Such homogenization methods employ a multiple length-scale analysis, with the spatial period furnishing one of the characteristic length scales. An analysis of suspension behavior has been made via such techniques by Lévy and Sanchez-Palencia (1983a,b) for dilute suspensions of particles and droplets, as well as for concentrated suspensions of solid particles. Nguetseng (1982) analyzed the behavior of phase mixtures subjected to vibrations.

This section begins with an account of spatially periodic suspension models embodying a single particle (a solid sphere in most cases) per unit cell. Rigidity of the suspended particles results in the existence of a maximum possible kinematic concentration whose numerical value is shown to depend explicitly on the macroscopic flow field being considered. Instantaneous, configuration-specific, suspension-scale rheological properties, amounting to a spatial average of the comparable microscale properties, follow from computation of the local interstitial dynamical fields. Time averages of these dynamical fields can be meaningfully performed when, in addition to the instantaneous spatial periodicity, the motion of the lattice is simultaneously time periodic. In such circumstances, the particulate microscale configuration of the suspension reproduces itself in time; it is over one such period of the self-reproducing lattice that the time average is to be performed. Several results are given for both dilute and concentrated suspensions. Finally, a unit cell containing N particles (with N possibly large) is considered, and some preliminary results for such systems are outlined.

A. Description and Kinematics of Spatially Periodic Suspensions

This subsection, which finds its genesis in the work of Adler and Brenner (1985a), is organized as follows: Following a brief introduction to the physical system under study, the kinematical possibility of the existence of a microscale interstitial fluid motion, consistent with a prescribed macroscale shearing motion of the suspension as a whole, is analyzed in the context of the mutual impenetrability of the suspended particles. This allows the concept of a "maximal kinematic concentration" to be defined. Next, the reproducibility (or, more generally, "near reproducibility") of the lattice in time is studied. Such temporal configurational periodicity is critical when time averaging of instantaneous quasistatic states needs to be performed to obtain the effective properties of the suspension.

Consider a suspension composed of an ordered, repetitive, three-dimensional array of identical, rigid spheres immersed in an otherwise homogeneous fluid continuum and extending indefinitely in every direction. From a formal point of view, the lattice Λ, representing the group of translational self-coincidence symmetry operations of this spatially periodic medium, consists of the set of points

$$\mathbf{R_n} = n_1 \mathbf{l}_1 + n_2 \mathbf{l}_2 + n_3 \mathbf{l}_3 \quad (n_j = 0, \pm 1, \pm 2, \ldots) \quad (j = 1, 2, 3). \quad (7.1)$$

Here, $(\mathbf{l}_1, \mathbf{l}_2, \mathbf{l}_3)$ denote three linearly independent vectors of $\mathbb{R}^3$, serving as a basis of Λ, whereas $\{n_1, n_2, n_3\} \equiv \mathbf{n}$, say, is a triad of integers. The symbol $\mathbf{0} \equiv \{0, 0, 0\}$ will be arbitrarily chosen to designate the lattice origin.

Equivalently, the lattice Λ may be represented by the second-order tensor

$$\mathbf{L} = l_1 \mathbf{e}_1 + l_2 \mathbf{e}_2 + l_3 \mathbf{e}_3, \tag{7.2}$$

with the trio of unit vectors $(\mathbf{e}_1, \mathbf{e}_2, \mathbf{e}_3)$ denoting an orthonormal basis of the vector space $\mathbb{R}^3$. The determinant $d(\Lambda)$ of the lattice, corresponding to the superficial volume of a unit cell, is defined as

$$d(\Lambda) = |\det \mathbf{L}|. \tag{7.3}$$

Consider the centers of the identical spherical particles of radii a to be instantaneously located at the lattice points $\mathbf{R}_n$. As such, the simplest geometric state exists, in which only one particle is contained within each unit cell. When the latter suspension is sheared, the three basic lattice vectors $\mathbf{l}_i$ ($i = 1, 2, 3$) (or, equivalently, the dyadic $\mathbf{L}$) become functions of time t. Under a homogeneous deformation, the lattice composed of the sphere centers remains spatially periodic, although its instantaneous spatially periodic configuration necessarily changes with time.

In the present rheological context, lattice deformation may be regarded as arising from the transport of neutrally buoyant lattice points suspended within a macroscopically homogeneous linear shear flow. The *local* vector velocity field $\mathbf{v}$ at a general (interstitial or particle interior) point $\mathbf{R}$ of such a spatially periodic suspension can be shown to be of the form

$$\mathbf{v}(\mathbf{R} + \mathbf{R}_n) - \mathbf{v}(\mathbf{R}) = \mathbf{R}_n \cdot \mathbf{G}, \tag{7.4}$$

with $\mathbf{G}$ as the constant macroscopic velocity gradient dyadic. $\mathbf{G}$ is assumed to be independent of both position and time. Hence, the local velocity gradient $\nabla \mathbf{v}$ is instantaneously spatially periodic for all time. For the applications that follow, it is further supposed that the strain rate $\mathbf{G}$ is traceless:

$$\operatorname{tr} \mathbf{G} = 0, \tag{7.5}$$

corresponding to the combination of an incompressible fluid, $\nabla \cdot \mathbf{v} = 0$, and rigid particles. The foregoing considerations, together with an assumed no-slip condition on the sphere surfaces, define the state of the system.

Since rigid spheres are nondeformable, it is known that in a *static* system, $d(\Lambda)$ possesses a minimum value (over all possible lattices) to which corresponds a maximal solids concentration

$$\delta = 4\pi a^3 / 3 d_{\min}(\Lambda). \tag{7.6}$$

This yields $\delta \approx 0.74$, a value first calculated by Gauss in 1831. Similar considerations can be extended to nonspherical (e.g., convex) particles; the corresponding mathematical framework constitutes a portion of the so-called geometry of numbers (Lekkerkerker, 1969).

Equivalent considerations for nonstatic, sheared systems demonstrate the kinematical possibility of such shearing motions. This requires, *inter alia*, that the distance between any two sphere centers remains larger than $2a$. The static viewpoint can be generalized to such circumstances as follows: Rather than considering the lattice deformation, it suffices to examine the deformed "collision sphere." The latter body $\mathfrak{S}$ is defined as the set of points

$$\mathfrak{S} = \{\mathbf{x} : \mathbf{x} = (\exp \mathbf{G}^\dagger t) \cdot \mathbf{x}_0, \quad \forall \|\mathbf{x}_0\| \leq 2a, \quad \forall t \in (-\infty, \infty)\}. \tag{7.7}$$

When the lattice points initially lie entirely outside $\mathfrak{S}$ (except $\mathbf{0}$, which lies at the center of $\mathfrak{S}$), it can be shown (Adler and Brenner, 1985a) that macroscopic shearing motion is possible for all times.

A minimal value of the determinant $d(\Lambda)$ corresponds to the body $\mathfrak{S}$, and thus, also corresponds to a maximum kinematic concentration, a quantity of some practical importance. Determination of the maximum kinematic concentration for a flow characterized by a given velocity gradient $\mathbf{G}$ requires the corresponding configuration of $\mathfrak{S}$ to be calculated and its minimal determinant to be deduced (generally from geometric number theory). Calculations have been performed for both two- and three-dimensional shearing motions (Adler, 1984). The most general two-dimensional incompressible linear flow can be written parametrically as (Kao *et al.*, 1977)

$$u = Gy, \quad v = \lambda Gx, \quad w = 0, \tag{7.8}$$

with (u, v, w) the respective (x, y, z) components of the velocity field, G the magnitude of the shear rate, and λ a nondimensional parameter lying in the

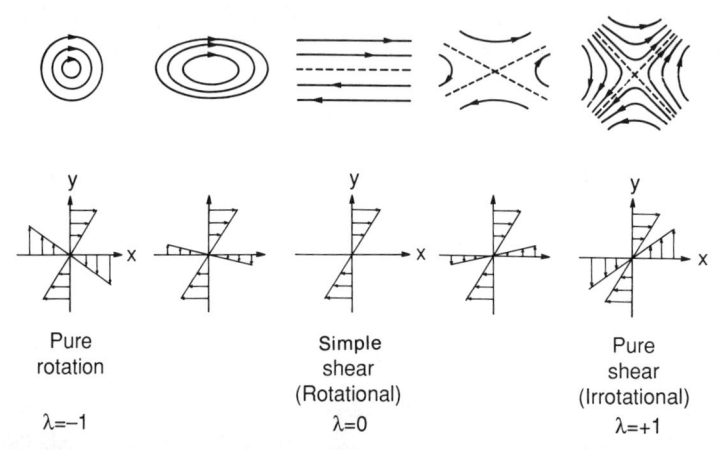

FIG. 1. Schematic of a family of two-dimensional steady incompressible shear flows showing the streamline patterns at the top and the corresponding velocity components at the bottom. By varying λ continuously from -1 to $+1$, the flow can be varied from pure rotation (without strain) to pure strain (without rotation).

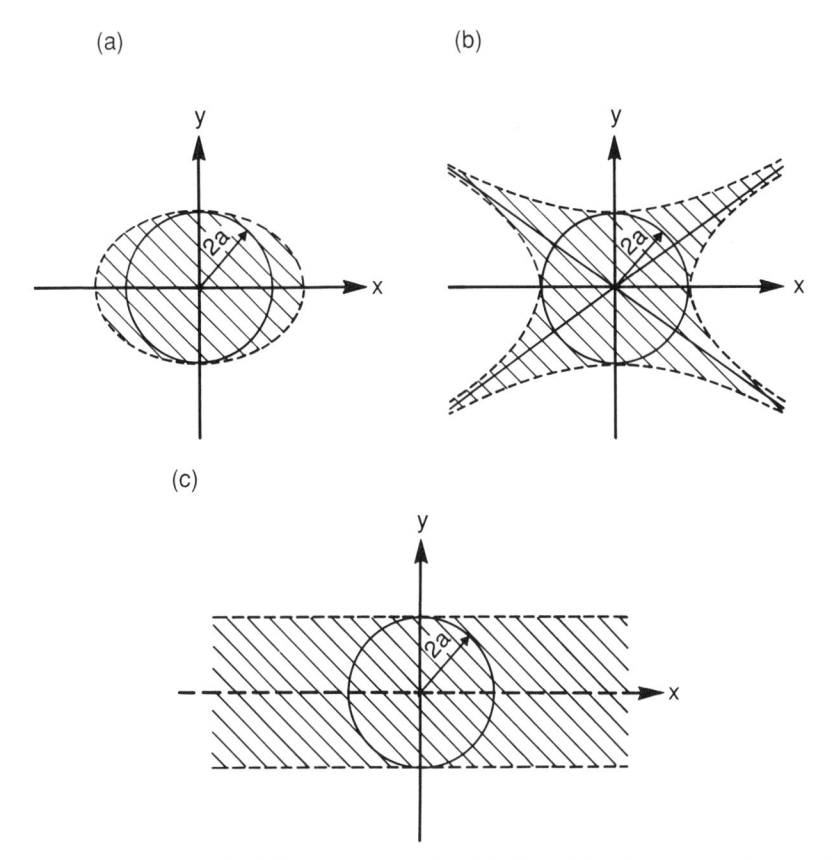

FIG. 2. The star body $\mathscr{S}$ for a two-dimensional lattice subjected to a two-dimensional shear flow. (a) Elliptic streamlines, $\lambda < 0$; (b) hyperbolic streamlines, $\lambda > 0$; (c) simple shear flow, $\lambda = 0$.

range $-1 \leq \lambda \leq 1$, encompassing all possible flow variants. Figure 1 (Kao et al., 1977) summarizes the entire family of trajectories.

The body $\mathfrak{S}$ is illustrated in Fig. 2 for the simple case of a spatially periodic suspension of circular disks subjected to the flow given by Eq. (7.8). It is evident from the figure that any point lying within the body $\mathfrak{S}$ will later be found within the collision disk, contradicting the specified impenetrability condition imposed on these rigid bodies.

A similar analysis can be performed for three-dimensional lattices subjected to the same flow. The corresponding maximum concentration curve in three dimensions is shown in Fig. 3 as a function of the flow parameter λ. This curve displays a discontinuous dependence on λ in the neighborhood of $\lambda = 0$, revealing a very special feature of simple shear flow. The "saw-tooth" property characterizing hyperbolic flows ($\lambda > 0$) is derived from the best estimates

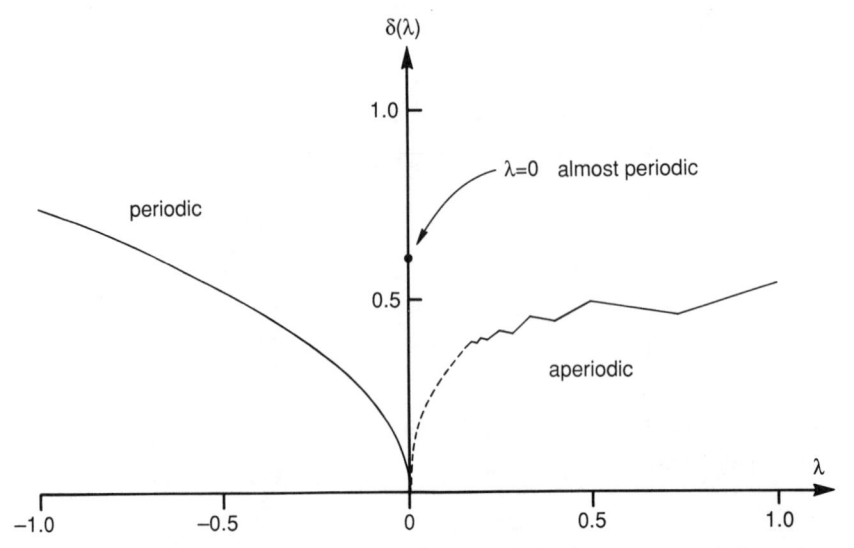

FIG. 3. The maximum density $\delta(\lambda)$ as a function of the flow parameter λ for a three-dimensional lattice.

currently available for the minimum value of $d(\Lambda)$, obtained from the geometry of numbers. Such estimates are not definitive in any absolute sense.

In summary, the key point of the foregoing analysis is the existence of a maximum particulate-phase concentration whose value depends explicitly on the specified shear flow in which the particles are suspended.

Finally, the self-reproducibility in time of the lattice configuration (for two-dimensional flows) must be addressed. In the elliptic streamline region ($\lambda < 0$), the lattice necessarily replicates itself periodically in time owing to closure of the streamlines. For hyperbolic flows ($\lambda > 0$), the lattice is not generally reproduced; however, in connection with research on spatially periodic models of foams (Aubert *et al.*, 1986; Kraynik, 1988), Kraynik and Hansen (1986, 1987) found a finite set of reproducible hexagonal lattices for the extensional flow case $\lambda = 1$. It is not clear how this unique discovery can be extended, if at all.

The most interesting case arises for simple shear flows ($\lambda = 0$). It can be proven mathematically that time-periodic motion is possible only when the direction of the flow is parallel to either a lattice plane or a lattice line. These are respectively termed "slide" and "tube" flows; the origin of such terminology is evident from Fig. 4. Very similar configurations have been experimentally observed by Ackerson and Clark (1983).

For simplicity of presentation, and since no novel results are furnished by tube flows, subsequent attention is restricted to slide flows. As in Fig. 4(a), the

RHEOLOGICAL MODELS OF SUSPENSIONS 43

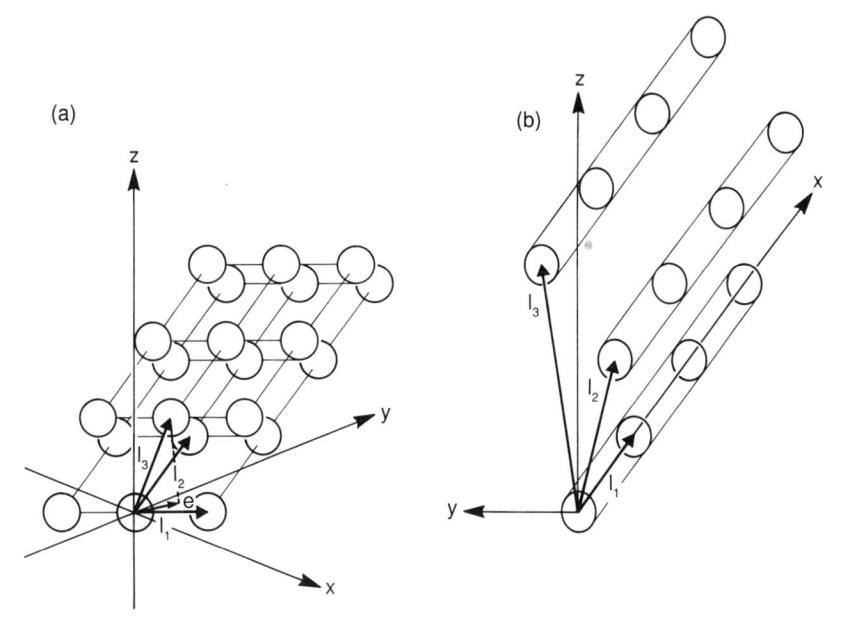

FIG. 4. The two possible configurations for a three-dimensional lattice in a simple shear flow. The velocity field is given by $u = Gz$; e is the projection of l_3 onto the x-y plane: (a) slide flow, (b) tube flow.

relative positions of two slide planes are determined by l_3. Since the projection of l_3 onto the z-axis remains constant in time, only its temporal projection e onto the x-y plane needs to be known. Inasmuch as that projection is a straight line, and since l_3 is only required modulo the spatial periods l_1 and l_2, determination of the reproducibility (or lack thereof) gives rise to a classical problem in the theory of numbers (Fig. 5). Thus, depending on the rational or

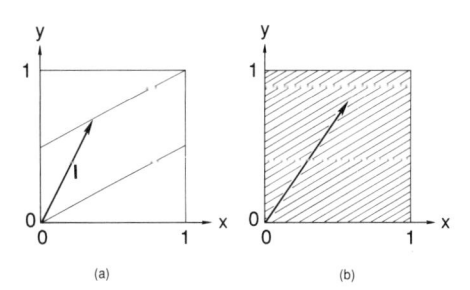

FIG. 5. Illustration of the interaction of the vector $e(t)$ with the unit square. (a) The slope is rational and equal to $\frac{1}{2}$; e visits the two segments in a periodic manner. (b) The slope is irrational; e visits the whole square with a uniform probability.

irrational character of the slope (cf. Fig. 5), the temporal lattice behavior will either be periodic or "almost periodic" in simple shear flow. For a rational slope, the terminus of the vector **e** visits a finite number of segments; otherwise, **e** visits the entire unit square $[0, 1]^2$ with uniform probability. Such knowledge proves essential when time averaging is to be performed.

B. Rheology

This subsection describes general dynamical properties of the interstitial velocity and pressure fields, from which the instantaneous rheological properties of the spatially periodic suspension are deduced, followed by the requisite time averaging for self-reproducing structures (Adler *et al.*, 1985). As a consequence of Eq. (7.4), the velocity field may be decomposed into respective linear (**R** · **G**) and spatially periodic ($\check{\mathbf{v}}(\mathbf{R})$) contributions. With account taken of the interstitial velocity vector $\bar{\mathbf{v}}^*$, the velocity field may be written as

$$\mathbf{v}(\mathbf{R}) = \mathbf{R} \cdot \mathbf{G} + \check{\mathbf{v}}(\mathbf{R}) + \bar{\mathbf{v}}^*, \tag{7.9}$$

where the spatially periodic field $\check{\mathbf{v}}(\mathbf{R})$ is assigned zero mean when averaged over the unit cell. The translational velocity $\mathbf{U_n}$ of particle **n** may be similarly decomposed as

$$\mathbf{U_n} = \mathbf{U_0} + \mathbf{R_n} \cdot \mathbf{G}, \tag{7.10}$$

with $\mathbf{U_0}$ the translational velocity of particle **0**.

Adherence of fluid to the particle surfaces requires satisfaction of the no-slip boundary conditions

$$\check{\mathbf{v}}(\mathbf{R_n} + \mathbf{r}) = (\mathbf{U_0} - \bar{\mathbf{v}}^*) + (\mathbf{\Omega} - \bar{\omega}) \times \mathbf{r} - \mathbf{r} \cdot \bar{\mathbf{S}} \quad \text{when } |\mathbf{r}| = a \tag{7.11}$$

with $\mathbf{\Omega}$ as the common angular velocity of each of the particles. Additionally, $\bar{\omega} = -\bar{\varepsilon}:\mathbf{G}/2 (\equiv -\varepsilon_{ijk}G_{kj}/2)$ is the pseudovector of the antisymmetric portion of **G**, and $\bar{\mathbf{S}} = (\mathbf{G} + \mathbf{G}^\dagger)/2$ is its corresponding symmetric portion.

Stokes Eqs. (2.2) and (2.3), in conjunction with Eq. (7.9), imply the spatial periodicity of the pressure gradient. As such, the pressure field may be decomposed as

$$p(\mathbf{R}) = \check{p}(\mathbf{R}) - \mathbf{R} \cdot \bar{\mathbf{F}}, \tag{7.12}$$

with $\check{p}(\mathbf{R})$ a spatially periodic field, and $\bar{\mathbf{F}}$ a constant vector equal to the force per unit volume exerted by the fluid on the particles.

Arguments much the same as those presented in Section II, together with the linear nature of the Stokes equations, yield the dynamical relationship

$$\begin{bmatrix} \mathbf{F} \\ \mathbf{N} \\ \mathbf{A} \end{bmatrix} = \mu_0 \begin{bmatrix} {}^t\mathbf{K} & \mathbf{C} & \overset{...}{\phi} \\ (\tau_f/\tau_0)\mathbf{C}^\dagger & {}^r\mathbf{K} & \overset{..}{\tau} \\ (\tau_f/\tau_0)\overset{...}{\mathbf{M}} & \overset{...}{\mathbf{N}}* & \overset{....}{\mathbf{Q}} \end{bmatrix} \begin{bmatrix} \bar{\mathbf{v}}^* - \mathbf{U}_0 \\ \bar{\omega} - \Omega \\ \bar{\mathbf{S}} \end{bmatrix}, \quad (7.13)$$

relating the hydrodynamic force $\mathbf{F}$, torque $\mathbf{N}$ (about the particle center), and particle stress $\mathbf{A}$ to the kinematical forcing terms appearing on the RHS. As a consequence of the Lorentz reciprocal theorem (Happel and Brenner, 1965), a number of dynamic symmetry relations (Adler et al., 1985) also apply to the coefficient matrix appearing in Eq. (7.13).

The partitioned "grand resistance matrix" in Eq. (7.13) is a function only of the instantaneous geometrical configuration of the particulate phase. This consists of the fixed particle shapes together with the variable relative particle positions and orientations. As such, geometrical symmetry arguments (where such symmetry exists) may be used to reduce the number of independent, nonzero scalar components of the coefficient tensors in Eq. (7.13) for particular choices of coordinate axes (e.g., "principal axis" systems).

Of special interest in applications is the important case of centrosymmetric particles, including spheres, ellipsoids, circular disks, and rods [cf. Brenner (1974) for a comprehensive account of the ensuing geometric symmetry simplifications]. Including the translational symmetry of the lattice, a suspension composed of individually centrosymmetric particles possessing identical orientations is itself centrally symmetric since an "empty" (i.e., particle-free) lattice necessarily possesses this point-group symmetry element. In such circumstances, geometric symmetry arguments (Brenner, 1974) reduce Eq. (7.13) to the form

$$\begin{bmatrix} \mathbf{F} \\ \mathbf{N} \\ \mathbf{A} \end{bmatrix} = \mu_0 \begin{bmatrix} {}^t\mathbf{K} & 0 & 0 \\ 0 & {}^r\mathbf{K} & \overset{..}{\tau} \\ 0 & \overset{...}{\mathbf{N}}* & \overset{....}{\mathbf{Q}} \end{bmatrix} \begin{bmatrix} \bar{\mathbf{v}}^* - \mathbf{U}_0 \\ \bar{\omega} - \Omega \\ \bar{\mathbf{S}} \end{bmatrix}. \quad (7.14)$$

This form significantly simplifies the algebraic structure of the kinetic problem. In particular, no direct coupling now exists between the translational motion and the angular and shearing motions, suggesting the possibility of treating the translational motion independently of the latter two.

In principle at least, Eq. (7.14) provides the basis for a complete calculation of the configuration-specific rheological properties of the suspension at each given instant of time t. In order to calculate therefrom the time-averaged properties of the suspension, consider the macroscopic simple shear flow discussed previously in connection with Fig. 4.

Consider a function f (of any tensorial rank) that is functionally dependent only on the geometry of the suspension—for example, in the case of spheres, on the geometry of the lattice $\Lambda(t)$. For simple shear flow, f will generally depend on both the static (time independent) and kinematic (time dependent) elements of the lattice. In order to simplify the representation, the static components will not appear explicitly among the arguments of f. Thus, notationally we write

$$f[\mathbf{L}(t)] = f[\mathbf{e}(t)], \qquad (7.15)$$

with $\mathbf{e}(t)$ as the projected lattice vector previously defined. Of course, when specific lattice calculations are actually performed, the static elements play a role too and need be incorporated.

The quantity of physical interest in applications normally proves to be not the instantaneous value of the function f itself, but rather its time average, defined as

$$\langle f \rangle = \lim_{T \to \infty} \frac{1}{T} \int_0^T f[\mathbf{e}(t)] \, dt. \qquad (7.16)$$

In order to effect this integration, the behavior of $\mathbf{e}(t)$ within the unit square must be known. In accordance with the preceding development, a distinction must be made between the respective two cases of rational and irrational values of the slope of $\mathbf{e}$.

When the slope ξ is rational [cf. Fig. 5(a)], $\mathbf{e}$ possesses a time-periodic trajectory $\mathscr{L}$ within the unit square of total length L. The required average [Eq. (7.16)] can then be computed as

$$\langle f \rangle = \frac{1}{L} \int_{\mathbf{e} \in \mathscr{L}} f(\mathbf{e}) \, dl, \qquad (7.17)$$

whose computation now appears in the guise of a purely geometric time-independent problem, where dl is a differential line element along the trajectory $\mathscr{L}$.

For irrational values of ξ, the terminus of the $\mathbf{e}$ vector visits all accessible positions in the unit square with uniform probability. Hence, the time average $\langle f \rangle$ can be shown (Adler et al., 1985) to be equivalent to the integral

$$\langle f \rangle = \int_{[0,1]^2} f(\mathbf{e}) \, dx \, dy \qquad (7.18)$$

over the unit square.

This generic formula permits averages to be calculated for any pertinent suspension property f. When time-independent external forces $\mathbf{F}^{(e)}$ and torques $\mathbf{N}^{(e)}$ act on each of the suspended particles, Eq. (7.14), together with the net force- and torque-free conditions characterizing the neutrally buoyant particles, yields

$$\langle \mathbf{U}_0 - \bar{\mathbf{v}}^* \rangle = \mu_0^{-1} \langle {}^t\mathbf{K}^{-1} \rangle \cdot \mathbf{F}^{(e)}$$
$$\langle \mathbf{\Omega} - \bar{\omega} \rangle = \langle {}^r\mathbf{K}^{-1} \cdot \ddot{\tau} \rangle : \bar{\mathbf{S}} + \mu_0^{-1} \langle {}^r\mathbf{K}^{-1} \rangle \cdot \mathbf{N}^{(e)} \quad (7.19)$$
$$\langle \mathbf{A} \rangle = -\langle \dddot{\mathbf{N}}^* \cdot {}^r\mathbf{K}^{-1} \rangle \cdot \mathbf{N}^{(e)} - \mu_0 \langle \dddot{\mathbf{N}}^* \cdot {}^r\mathbf{K}^{-1} \cdot \ddot{\tau} - \dddot{\mathbf{Q}} \rangle : \bar{\mathbf{S}}.$$

Preceding results for the simple shear case are potentially relevant for a variety of applications in the sense that all stationary suspension-scale rheological properties of interest may now be derived in a rigorous straightforward (albeit tedious) manner. The essentially geometric character of these dynamical results again bears emphasis. This is, perhaps, not surprising in retrospect since the particle kinematics have been reduced to a geometric quadrature, while the efficacy of the grand resistance matrix (which now embodies the dynamical elements of the problem) stems from its dependence on only the geometrical configuration of the particulate phase. Obviously, this pair of independent geometrical elements supplement one another nicely; together they constitute a potent combination useful, perhaps, in even more general circumstances.

A second prominent feature here is the ergodic character (or lack thereof) of the process, depending on the rationality or irrationality of ξ. This leads inevitably to the fascinating question, "Does a real system choose between these values of ξ, and if so, how?" The boundaries themselves remain neutral with respect to the choice of ξ whenever they are compatible with the flow. Thus, for a slide flow, the walls must be parallel to the slides, whereas for a tube flow, they must be parallel to the tube. In both cases there remains an additional degree of freedom, which is precisely the choice of ξ. Other examples of indeterminancy arise from the neglect of fluid and particle inertia, as already discussed in Section I (see also the review in Leal, 1980). Whether or not inclusion of nonlinear inertial effects can remove the above indeterminacy, as it often does for the purely hydrodynamic portion of the problem, is a question that lies beyond the scope of the present (linear) Stokesian context.

Finally, hyperbolic and elliptic flows deserve at least a few comments. A suspension undergoing a hyperbolic flow does not generally reproduce itself in time. Accordingly, its time average is not physically meaningful, even if it is assumed to exist. Elliptic flow represents perhaps the simplest case among the class of two-dimensional flows since it is always self-reproducing as a consequence of the closed streamlines. However, its very simplicity is itself a source of disappointment since it is a relatively straightforward matter to determine the configurations explored by the suspension during each period and perform the requisite time integrations along these configurations. Thus, our geometric scheme offers no advantages over this direct temporal approach.

C. Cubic Arrays

This subsection summarizes available numerical results for the rheological properties of cubic arrays of neutrally buoyant spherical particles suspended in shear flows. [Comparable porous media calculations for seepage flows through fixed spatially periodic rectangular arrays of cylinders and cubic arrangements of spheres, encompassing the complete range of particle concentrations, are presented by Sangani and Acrivos (1982).]

A general solution of Stokes equations can be obtained by analytic continuation of the interstitial velocity and pressure fields into the interior of the regions occupied by the spheres, replacing the particle interiors by singular multipole force distributions located at the sphere centers $\mathbf{R_n}$ (Zuzovsky et al., 1983). Explicitly, $(\mathbf{v}, p)$ satisfies the dynamical equation

$$\mu_0 \nabla^2 \mathbf{v} = \nabla p - \sum_{\alpha=0}^{\infty} \sum_{\mathbf{n}} \nabla^\alpha \delta(\mathbf{R} - \mathbf{R_n})(\cdot)^\alpha \mathbf{F}_{\alpha+1}, \tag{7.20}$$

where $\nabla^\alpha = \nabla, \ldots, \nabla$ denotes α successive applications of the gradient operator; $\delta(\mathbf{R} - \mathbf{R_n})$ is the Dirac delta function, whereas $\mathbf{F}_{\alpha+1}$ is a completely symmetric constant tensor of rank $\alpha + 1$ to be determined. The symbol $(\cdot)^\alpha$ represents α successive dot multiplications; additionally,

$$\sum_{\mathbf{n}} \equiv \sum_{n_1=-\infty}^{\infty} \sum_{n_2=-\infty}^{\infty} \sum_{n_3=-\infty}^{\infty}. \tag{7.21}$$

Equation (7.20) is to be regarded as applying throughout all of space, including the particle interiors. Boundary conditions imposed at the sphere surfaces implicitly furnish the relations determining the $\mathbf{F}$'s, these being related to the forces, torques, etc., exerted on the fluid by the particles. For instance, the vector $\mathbf{F}_1$ is related to the external force acting on each particle via the expression

$$\mathbf{F}_1 = -\mathbf{F}_{\text{ext}}, \tag{7.22}$$

whereas $\mathbf{F}_2$ is expressible as

$$\mathbf{F}_2 = \mathbf{A} + \tau_0 \bar{\mathbf{P}}^a, \tag{7.23}$$

with $\mathbf{A}$ the so-called particle stress, $\bar{\mathbf{P}}^a$, the antisymmetric portion of the macroscopic stress tensor $\bar{\mathbf{P}}$, and τ_0 the superficial volume of a unit cell.

Calculations have thus far been performed for the three standard cubic arrays, namely simple cubic (sc), body-centered cubic (bcc), and face-centered cubic (fcc). As a result of this geometric symmetry, the couple $\mathbf{N}$ and particle stress dyadic $\mathbf{A}$ are given by the configuration-specific relations

$$\begin{aligned}\mathbf{N} &= \mu_0 \xi^{(r)}(\mathbf{\Omega}) - \bar{\omega}) \\ \mathbf{A} &= \mu_0 \ddddot{\psi}^{(s)} : \bar{\mathbf{S}}.\end{aligned} \tag{7.24}$$

TABLE II

COEFFICIENTS FOR USE IN EQS. (7.26)

Lattice type	n	$\tilde{a}_{20}$	$\tilde{b}_{20}$
sc	1	0.2857	−0.04655
bcc	2	−0.0897	0.01432
fcc	4	−0.0685	0.01271

Upon appropriate reduction in the number and nature of the independent tensorial components of $\overset{...}{\psi}{}^{(s)}$ ($\equiv \psi^{(s)}_{ijkl}$), resulting from the common point-group symmetry elements of the sphere and cube (applied to fourth-rank tensors), the material tensor can be shown quite generally to be of the form (Zuzovsky et al., 1983)

$$\overset{...}{\psi}{}^s/\tau_0 = \alpha(\mathbf{e}_1\mathbf{e}_1\mathbf{e}_1\mathbf{e}_1 + \mathbf{e}_2\mathbf{e}_2\mathbf{e}_2\mathbf{e}_2 + \mathbf{e}_3\mathbf{e}_3\mathbf{e}_3\mathbf{e}_3)$$
$$+ \beta[(\mathbf{e}_1\mathbf{e}_2 + \mathbf{e}_2\mathbf{e}_1)(\mathbf{e}_1\mathbf{e}_2 + \mathbf{e}_2\mathbf{e}_1)$$
$$+ (\mathbf{e}_2\mathbf{e}_3 + \mathbf{e}_3\mathbf{e}_2)(\mathbf{e}_2\mathbf{e}_3 + \mathbf{e}_3\mathbf{e}_2)$$
$$+ (\mathbf{e}_3\mathbf{e}_1 + \mathbf{e}_1\mathbf{e}_3)(\mathbf{e}_3\mathbf{e}_1 + (\mathbf{e}_1\mathbf{e}_3)]. \quad (7.25)$$

Coefficients α, β, and $\xi^{(r)}$ are determined by solving Stokes equations.

For dilute arrays these coefficients are given by the expressions

$$\alpha = (5/2)\phi[1 - (1 - 60\tilde{b}_{20})\phi + 12\tilde{a}_{20}\phi^{5/3} + 0(\phi^{7/3})]^{-1}$$
$$\beta = (5/2)\phi[1 - (1 + 40\tilde{b}_{20})\phi - 8\tilde{a}_{20}\phi^{5/3} + 0(\phi^{7/3})]^{-1} \quad (7.26)$$
$$\xi^{(r)} = 6\phi[1 - \phi + 12\tilde{a}^2_{20}\phi^{10/3} + 0(\phi^{14/3})]^{-1}.$$

Requisite coefficients $\tilde{a}_{20}$ and $\tilde{b}_{20}$ are tabulated in Table II for each of the three cubic lattices. These values agree well with those independently derived by Kapral and Bedeaux (1978).

For highly concentrated suspensions where the spheres almost touch, these coefficients adopt the asymptotic forms (Zuzovsky et al., 1983)

$$\alpha = (3\pi/16)\chi^{1/3}(1 - \chi^{1/3})^{-1},$$
$$\beta = (\pi/4)\ln(\chi^{-1/3} - 1)^{-1}, \quad (7.27)$$
$$\xi^{(r)} = \pi\ln(\chi^{-1/3} - 1)^{-1},$$

for simple cubic arrays. Here, $\chi = \phi/\phi_{\max}$, with $\phi_{\max} = \pi/6$ for a simple cubic array.

Equations (7.26) and (7.27) furnish limiting results for both dilute and concentrated systems. Nunan and Keller (1984) subsequently extended these numerical calculations to intermediate concentrations as well, simultaneously confirming the dilute and concentrated sphere suspension asymptotes in Eqs. (7.26) and (7.27).

The foregoing configuration-specific rheological results are valid only instantaneously at the instant the sheared suspension momentarily possesses the specified cubic arrangement. As such, the configuration-specific rheological results will not be compared to formulas previously cited in Section II. In order to derive stationary values for the macroscopic rheological properties of suspensions, a large number of ordered structures must be analyzed and time averages [or the equivalent averages in Eq. (7.17) or (7.18)] computed therefrom. Thus, the average of any desired interstitial field property can clearly be derived or (more practically) estimated by interpolation. [This is reminiscent of the determination of the relative trajectories of two spheres in linear shear flows (Adler, 1981a,b).]

To conclude this subsection, we expose an interesting paradox arising from the time dependence of the particle configuration. As discussed in Section III, Frankel and Acrivos (1967) developed a time-independent "lubrication" model for treating concentrated suspensions. Their result, given by Eq. (3.7), predicts singular behavior of the shear viscosity in the maximum concentration limit where the spheres touch. Within the spatially periodic framework, the instantaneous macroscopic stress tensor may be calculated for the lubrication limit, $\varepsilon \to 0$. The symmetric portion of its deviatoric component takes the form (Zuzovsky et al., 1983)

$$\frac{\mu_0 \pi a}{2\tau_0 \varepsilon} \mathbf{mm} : \bar{\mathbf{S}} \left(3 \frac{\mathbf{mm}}{|\mathbf{m}|^2} - \mathbf{I} \right) \tag{7.28}$$

for the configuration shown in Fig. 6. Here, ε denotes the gap width made dimensionless with the sphere radius. The time integral in Eq. (7.17) resulting from integration of Eq. (7.28) can be shown (Zuzovsky et al., 1983) to be nonsingular! That is, terms of $O(1)$ result upon integration rather than terms that tend to infinity in the $\varepsilon \to 0$ limit. Thus, although the instantaneous stress tends to infinity in the touching-sphere limit, the time-average stress nevertheless remains finite—a fact that contradicts Eq. (3.7). These conclusions regarding the nonsingular nature of the touching-sphere limit are supported by the scaling arguments of Marrucci and Denn (1985) with respect to the important role played by the time-dependence of the relative sphere positions, in which arguments are not limited to periodic arrays.

Of course, in this limit, nonhydrodynamic factors may dominate the suspension's rheological properties. Included in this class are such potentially relevant features as surface roughness, interstitial fluid cavitation (Goldman et al., 1967a), particle elasticity and lattice disorder, any of which, when incorporated into the analysis, might give rise to singular behavior in the $\varepsilon \to 0$

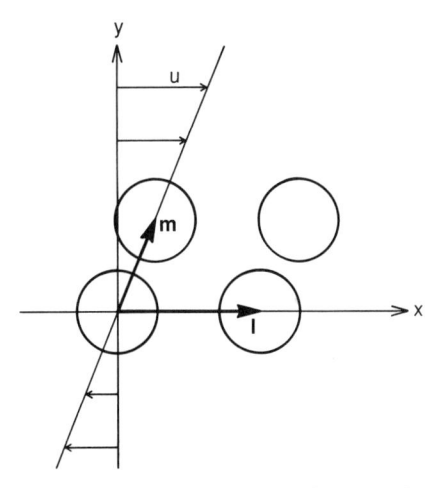

FIG. 6. Two-dimensional array in a simple shear flow; $u = Gy$, $v = 0$ are the velocity components.

limit. However, a detailed discussion of the relative importance of such phenomena in the interpretation and rationalization of experimental results appears premature at this stage. Comments advanced in Section VI regarding the possible formation of infinite clusters may also be relevant to the resolution of this paradox.

D. Extension to N Particles in a Unit Cell

The previous analysis may be extended to spatially periodic suspensions whose basic unit cell contains not one, but many particles. Such models would parallel those employed in liquid-state theories, which are widely used in computer simulations of molecular behavior (Hansen and McDonald, 1976). This subsection briefly addresses this extension, showing how the trajectories of each of the particles (modulo the unit cell) can be calculated and time-average particle stresses derived subsequently therefrom. This provides a natural entrée into recent dynamic simulations of suspensions, which are reviewed later in Section VIII.

Consider a spatially periodic suspension whose basic unit cell contains N particles, all of whose sizes, shapes, and orientations may be different. The particulate phase of the system is completely specified geometrically by the values of the $3N$ spatial coordinates $\mathbf{r}^N$, and $3N$ orientational coordinates $\mathbf{e}^N$ of the (generally nonspherical) particles, as in Section II,A. A particle is identified by the pair of scalar and "vector" indices i ($i = 1, \ldots, N$) and $\mathbf{n}$, respectively, the latter of which indicates the particular unit cell to which particle i is assigned.

Stokes Eqs. (2.2) and (2.3) are again assumed to describe the interstitial fluid dynamics and kinematics, while the adherence condition at the particle surfaces now requires that

$$\mathbf{v} = \mathbf{U}_{i,\mathbf{n}} + \mathbf{\Omega}_i \times \mathbf{r}_i \quad \text{on } s_i, \tag{7.29}$$

with $\mathbf{U}_{i,\mathbf{n}}$ and $\mathbf{\Omega}_i$ the respective translational and rotational velocities of particle $\{i, \mathbf{n}\}$. Note that $\mathbf{\Omega}_i$ does not depend on the cell number $\mathbf{n}$ for the homogeneous macroscopic shear flows being considered. The velocity of the particle $\{i, \mathbf{n}\}$ can be decomposed, as in Eq. (7.13), into the sum

$$\mathbf{U}_{i,\mathbf{n}} = \mathbf{U}_i + \mathbf{R}_\mathbf{n} \cdot \mathbf{G}, \tag{7.30}$$

with $\mathbf{U}_i$ the velocity of particle i in cell $\mathbf{0}$. Upon employing the decomposition in Eq. (7.9), which continues to be valid in present circumstances, the counterpart of Eq. (7.11) here adopts the form

$$\check{\mathbf{v}} = (\mathbf{U}_i - \bar{\mathbf{v}}^*) + (\mathbf{\Omega}_i - \bar{\omega}) \times \mathbf{r}_i - \mathbf{r}_i \cdot \bar{\mathbf{S}}, \quad \text{on } s_i. \tag{7.31}$$

As a consequence of the linearity of the governing equations, the grand resistance matrix formulation is again applicable, yielding

$$\begin{bmatrix} \mathbf{F}^N \\ \mathbf{N}^N \\ \mathbf{A} \end{bmatrix} = \mu_0 \mathbf{R}(\mathbf{r}^N, \mathbf{e}^N, \mathbf{L}) \begin{bmatrix} \bar{\mathbf{v}}^{*N} - \mathbf{U}^N \\ \bar{\omega}^N - \mathbf{\Omega}^N \\ \bar{\mathbf{S}} \end{bmatrix}, \tag{7.32}$$

where the partitioned matrix $\mathbf{R}$ is a material property dependent only on the configuration $(\mathbf{r}^N, \mathbf{e}^N)$ of the particles and the configuration of the time-dependent lattice $\mathbf{L}$.

Upon neglecting particle inertial effects, the dynamics of the individual particle motions is governed by the force/torque balance

$$\begin{bmatrix} \mathbf{F}^N \\ \mathbf{N}^N \end{bmatrix} + \begin{bmatrix} \mathbf{F}^N_{\text{ext}} \\ \mathbf{N}^N_{\text{ext}} \end{bmatrix} = \begin{bmatrix} 0 \\ 0 \end{bmatrix}, \tag{7.33}$$

where $\mathbf{F}^N_{\text{ext}}$ and $\mathbf{N}^N_{\text{ext}}$ are each $3N$ vectors composed of the external three-dimensional force and torque vectors, if any, acting on the particles. Equation (7.33) together with Eq. (7.32) provides a first-order evolution system for $(\mathbf{r}^N, \mathbf{e}^N)$. Thus, as in Section II,A, for a suspension of spheres, a trajectory equation of the form

$$d\mathbf{r}^N/dt = \mathbf{f}(\mathbf{r}^N, \mathbf{L}) \tag{7.34}$$

can be obtained, with $\mathbf{f}$ as a nonlinear vector function. The trajectories $\mathbf{r}^N(t)$ of the N spheres may thus be determined, at least in principle, by solving Eq. (7.34).

Computation of the grand resistance matrix $\mathbf{R}$ for each possible particle configuration provides a major numerical challenge (see Section VIII). The various methods cited in Section II for dealing with the many-body problem are potentially useful in this context. Detailed calculations must be performed for a number of accessible configurations, and the configurational evolution determined by interpolation in order to effect the requisite time averaging.

Determination of the maximum particle concentration is also of interest since it no longer constitutes a purely kinematical problem. Rather, the suspension contained within the unit cell is now a mixed "object" possessing both solid-like and liquid-like features. In particular, it behaves like a solid insofar as mutual impenetrability demands are concerned, whereas it behaves like a liquid in its ability to change its configuration (i.e., it can "flow").

The question of ergodicity of the dynamical system in Eq. (7.34) remains. When the lattice exactly reproduces itself in time (as was the case for rational simple shear flow), $\mathbf{L}$, and hence $\mathbf{r}^N$, are periodic functions of time, whence the process is nonergodic. When $\mathbf{L}$ is almost periodic in time, so is $\mathbf{r}^N$. But it is not known whether $\mathbf{r}^N$ visits the entire available space or just some subspace thereof. This question must be answered before meaningful averages can be performed.

Another remaining problem arises when the results obtained for a finite cell are extrapolated to an infinite suspension. This necessitates the cell size to be much larger than any other length scales characterizing the suspension. While this is usually the case, it fails to be true near the "percolation threshold." Possible existence of the latter critical point is discussed by DeGennes (1979); if it does exist, a special technique called finite-size scaling (Stauffer, 1985) may be used to extract the asymptotic limit from numerical data gleaned on finite cells. This is equivalent to extracting the limiting behavior of an infinite suspension from the corresponding results derived for a finite suspension of N particles.

In conclusion, we note that the formulation of Eqs. (7.32) and (7.33) is formally complete, encompassing the possibilities of dissimilar nonspherical particles within a unit cell, interparticle forces, Brownian motion, etc. At the same time, the scheme is rigorous, at least within the formal constrained geometric framework adopted. Just as "one particle per cell" results have contributed to the identification of the key features of the suspension problem, so it is to be expected that N-particle dynamical analyses will provide more general answers to comparable long-standing questions pertaining to the rheological properties of concentrated suspensions.

VIII. Current Research Topics

Four novel approaches to contemporary studies of suspensions are briefly reviewed in this final section. Addressed first is Stokesian dynamics, a newly developed simulation technique. Surveyed next is a recent application of generalized Taylor dispersion theory (Brenner, 1980a, 1982) to the study of momentum transport in suspensions. Third, a synopsis is provided of recent studies in the general area of fractal suspensions. Finally, some novel properties (e.g., the existence of antisymmetric stresses) of dipolar suspensions are reviewed in relation to their applications to magnetic and electrorheological fluid properties.

A. STOKESIAN DYNAMICS

The extension to multiparticle unit cells alluded to in Section VII,E has, in some sense, already been achieved. Contributions by Bossis and Brady (1984) and Brady and Bossis (1985) provide a general technique for simulating the temporal evolution of sheared suspensions. These authors subsequently applied their technique to calculating the rheological properties of nondilute suspensions. This periodic boundary-condition scheme for simulating unbounded suspensions is closely related to that of Section VII,E, though we retain here the original notation of Bossis and Brady for ease of reference. In the paragraphs that follow, we content ourselves with merely reviewing the rudiments of this new simulation technique. A much more comprehensive survey has been provided by Brady and Bossis (1988), who present examples, of Stokesian dynamics applications to suspension rheology, porous media, diffusion of interacting particles, and bounded suspensions.

In the absence of inertia and Brownian motion, the quasistatic dynamics of N rigid suspended particles contained within a unit cell is governed by the composite force and torque balance

$$\mathbf{F}_H + \mathbf{F}_P = \mathbf{0}, \tag{8.1}$$

in which $\mathbf{F}_H$ and $\mathbf{F}_P$ are $6N$ force–torque vectors of hydrodynamic and nonhydrodynamic origins, respectively. The equivalent grand resistance formulation [cf. Eq. (7.32)] for the hydrodynamic force–torque vector is given as

$$\mathbf{F}_H = \mathbf{R} \cdot \mathbf{U}^* + \overset{\dots}{\Phi}:\mathbf{E}, \tag{8.2}$$

with the particulate material entities $\mathbf{R}$ and Φ respectively referred to by Brady and Bossis (following Brenner and O'Neill, 1972) as the grand resistance and shear resistance matrices. $\mathbf{U}^*$ denotes the translational-angular velocity "vector" of the particles relative to the bulk, and $\mathbf{E}$ denotes the symmetric portion of the bulk rate of strain tensor (corresponding to $\bar{\mathbf{S}}$ in Section VII). Substitution of Eq. (8.2) into Eq. (8.1) and subsequent multiplication by $\mathbf{R}^{-1}$ yields the evolution equation

$$\mathbf{U}^* = -\mathbf{R}^{-1} \cdot \{\dddot{\boldsymbol{\Phi}}:\mathbf{E} + \mathbf{F_p}\}. \tag{8.3}$$

The simulation technique involves temporal integration of Eq. (8.3) to find the instantaneous positions and orientations of the suspended particles. Since $\mathbf{R}$ and $\dddot{\boldsymbol{\Phi}}$ are configuration dependent, they must be updated at each time step as the configuration evolves in time. (Allowance can also be made for the externally imposed rate-of-strain dyadic $\mathbf{E}$ to depend on time if desired, e.g., time periodicity.) Integration is continued beyond the time at which a "stationary" particle state is achieved, as characterized by monitoring the average of the square of the particle velocities relative to the bulk. (Note that questions of reproducibility of the structure or of its ergodicity are not addressed, although the computational results are themselves reproducible beginning from different initial configuration.)

This general technique was employed to simulate the motion of a monolayer of identical spherical particles subjected to a simple shear. Confinement to a monolayer represents tremendous economies of computer time compared with a three-dimensional simulation. Hopefully, these highly specialized monolayer results will provide comparable insights into the physics of three-dimensional suspensions. Periodic boundary conditions were used in the simulation, and the method of Evans (1979) was incorporated to reproduce the imposed shear.

In order to account for hydrodynamic interactions among the suspended particles, Bossis and Brady (1984) used both pairwise additivity of velocities (mobilities) and forces (resistances), discussing the advantages and disadvantages of each method. While their original work did not take explicit account of three- (or more) body effects, the recent formulation of Durlofsky, Brady, and Bossis (1988) does provide a useful procedure for incorporating both the far-field, many-body interactions and near-field, "lubrication" forces into the grand resistance and mobility matrices.

Simulations were carried out with and without the presence of additional nonhydrodynamic, interparticle forces. When included, the latter were assumed to be repulsive in nature and given pairwise for spheres by the DLVO-type expression

$$\mathbf{F}_{\text{rep}} = \mathbf{F}_0 \frac{\tau e^{-\tau h}}{1 - e^{-\tau h}}, \tag{8.4}$$

with $\mathbf{F}_0$, the amplitude of the force; τ, the inverse of the Debye length (made dimensionless with the sphere radius); and h, the dimensionless separation distance between adjacent sphere surfaces (Takamura et al., 1981; Bossis and Brady, 1984). Important parameters were the volumetric solids fraction ϕ, characteristic interaction length τ^{-1}, and shear rate $\dot{\gamma}$ (which is made dimensionless with $|F_0|/6\pi\mu a^2$).

Principal results pertaining to the long-time mean structure of the suspension calculated by Bossis and Brady (1984) are embodied in the pair distribution function $g(\mathbf{r}) = g(r, \vartheta)$, defined as the conditional probability density for finding a particle at $\mathbf{r}$, given the existence of another particle at the origin (and normalized by dividing by the particle number density). Angular structure is found to exist only in the presence of repulsive forces where a larger particle density exists on the upstream side of the sphere situated at the origin. (On the downstream side, both the repulsive forces and shear tend to separate the particles.) Radial structure is also found (mainly due to excluded volume effects), possessing classical nearest neighbor peaks, second nearest-neighbor peaks, etc. Both the angular and radial structures were found to be functionally dependent on the particle-number density, shear rate, and repulsive-force range. Increasing the areal fraction ϕ of the particles and/or the length range τ^{-1} of the DLVO-type repulsive forces causes a transition to a layered structure, in which lines of particles "slide" relative to one another. As densities approaching the maximum possible for flow are achieved, cluster formation is observed to occur. Experiments by Husband and Gadala-Maria (1987) have verified the formation of anisotropic structures in noncolloidal suspensions.

In addition to the microstructural geometrical features described above, macroscopic, dynamical, rheological properties of the suspensions are derived by Brady and Bossis (1985). Dual calculations are again performed, respectively with and without DLVO-type forces. When such forces are present, an additional contribution (the so-called "elastic" stress) to the bulk stress tensor exists. In such circumstances, the term (Batchelor, 1977; Brady and Bossis, 1985)

$$\frac{1}{V} \sum_{i=2}^{N} \sum_{j<i} \mathbf{r}^{ij} \mathbf{F}^{ij}, \tag{8.5}$$

must be added to the RHS of Eq. (2.3a), where $\mathbf{r}^{ij}$ is the center-to-center separation of particles i and j, and $\mathbf{F}^{ij}$ is the pairwise interparticle force acting between them.

With the bulk, deviatoric stress tensor denoted by σ, Bossis and Brady define the relative viscosity η_r of the suspension as

$$\langle \sigma_{xy} \rangle = \eta_r \eta_0 \dot{\gamma}. \tag{8.6}$$

Additionally, primary and secondary normal stress coefficients ψ_1 and ψ_2 are defined by the respective relations

$$\langle \sigma_{xx} \rangle - \langle \sigma_{yy} \rangle = -\psi_1 \dot{\gamma}^2,$$
$$\langle \sigma_{yy} \rangle - \langle \sigma_{zz} \rangle = -\psi_2 \dot{\gamma}^2. \tag{8.7}$$

Expressions are derived for these coefficients in terms of dynamical quantities (such as the relative radial and tangential velocities of pairs of spheres) directly calculable by the simulation scheme.

Normal stress differences do not exist in the absence of interparticle forces. Moreover, the relative viscosity of the suspension is a function of only ϕ. At particle densities approaching the maximum possible that still allow the suspension to flow, cluster size (and, as a result, the viscosity of the two-dimensional monolayer) appears to scale as

$$[1 - (\phi/\phi_{\max})^{1/2}]^{-1}, \tag{8.8}$$

with the $\frac{1}{2}$ exponent replaced by $\frac{1}{3}$ in three dimensions. However, since the density range investigated was limited, the possible emergence of a different mode of behavior at densities close to the "percolation threshold," $\phi_{\max}$, is not ruled out. Comments are also offered by Bossis and Brady on the finite limit obtained for the viscosity of a spatially periodic suspension (cf. Section VI and VII,B). Comparison of their results with experiments has been found to be satisfactory (Brady and Bossis, 1985), although this comparison involves an ad hoc rescaling to convert their areal fraction to the comparable volumetric fraction prevailing in the three-dimensional experiments.

Repulsive interparticle forces cause the suspension to manifest non-Newtonian behavior. Detailed calculations reveal that the primary normal stress coefficient ψ_1 [cf. Eq. (8.7)] decreases like $\dot{\gamma}^{-1}$. In contrast, the suspension viscosity displays shear-thickening behavior. This feature is again attributed to the enhanced formation of clusters at higher shear rates.

Similar methods were employed by Schonberg et al. (1986) to investigate the multiparticle motions of a finite collection of neutrally buoyant spheres suspended in a Poiseuille flow. They were also used by Ansell and Dickinson (1986) to simulate the fragmentation of a large colloidal floc in a simple shear flow.

Finally, we direct attention to Barnes et al. (1987), for their extensive review of applications of computer simulations to dense suspension rheology, and also to Hassonjee et al. (1988), for their numerical scheme for dealing with large clusters of spherical particles.

B. Momentum Tracer Methods

In a companion pair of contributions, Mauri and Brenner (1991a,b) introduce a novel scheme for determining the rheological properties of suspensions. Their approach extends generalized Taylor–Aris dispersion-theory moment techniques (Brenner, 1980a, 1982)—particularly as earlier addressed to the study of tracer dispersion in immobile, spatially periodic media (Brenner, 1980b; Brenner and Adler, 1982)—from the realm of *material* tracer transport in suspensions to that of *momentum* tracer transport. Among other things, the configuration-specific viscosity of a suspension *at rest* (i.e., at zero shear rate) is shown by Mauri and Brenner to be the same as that for a *flowing* suspension possessing the same instantaneous particle configuration, at least in the limit of small Reynolds numbers.

The underlying *ansatz* consists of deliberately introducing a weak*, low-intensity, impulsive source of momentum (a so-called momentum "tracer") into the interstices of an otherwise quiescent suspension and monitoring the temporal spread of this momentum tracer from its initial point of introduction as it "diffuses" throughout the suspension in consequence of the kinematic viscosity of the interstitial fluid. Central to interpreting the instantaneous moments of the resulting momentum density-distribution is the fact that the momentum originally introduced impulsively into the system is conserved, since no external (or inertial) forces act on the neutrally buoyant suspension. In an abstract sense, momental diffusion of this conserved momentum tracer occurs by a mechanism conceptually no different from that underlying the mass diffusion of a conserved material tracer, except that the rate of momental diffusion is governed by the kinematic viscosity $v = \mu_0/\rho$ of the interstitial fluid, whereas the rate of mass diffusion is governed by the molecular diffusivity D of the material tracer through the interstitial fluid. (Indeed, this is the basis for all analogies between mass and momentum transport in *homogeneous* fluids.) Consequently, by analogy, just as the long-time mean-square displacement of a Brownian material tracer, through the interstices of the suspension, provides the suspension-scale (mean) dispersivity dyadic $\bar{D}_{ij}$ (Brenner, 1980b; Brenner and Adler, 1982) in terms of the interstitial molecular diffusivity D and particulate-phase configuration, so too does the long-time spread of the "Brownian" momentum tracer in Mauri and Brenner's (1991a,b) theory furnish the suspension-scale kinematic viscosity tetradic $\bar{v}_{ijkl}$ in terms of the kinematic viscosity v of the interstitial fluid and this same particulate configuration.

Specifics of this approach are outlined in the next few paragraphs. Consider a spatially periodic "suspension," one whose density ρ is everywhere constant and whose kinematic viscosity $v(\mathbf{r})$ is everywhere a spatially periodic function of position, albeit perhaps discontinuous. (Generalization to the case of

---

* Such "weakness" assures that the initial arrangement of particles prior to introduction of the tracer is not sensibly disturbed by the impulse as it "diffuses" throughout the suspension. Thus, in contrast with *flowing* systems, the mode of particle arrangement is decoupled from the hydrodynamics of the momentum transport process (as well as being time independent). Hence, the geometry of the particulate phase may be arbitrarily specified, leading to the unambiguous concept of configuration-specific rheological properties as experimentally realizable, macroscopic, dynamic attributes of such configurations.

spatially periodic distributions of *rigid* particles is addressed subsequently.) The vector **r** arises upon decomposition of the position vector **R** into the sum

$$\mathbf{R} = \mathbf{R}_n + \mathbf{r} \tag{8.9}$$

of discrete [cf. Eq. (7.1)] and continuous vectors. In particular, the vector **r** is "local" (Brenner, 1980b) in nature, being defined only within the interior domain τ_0 of a unit cell (the magnitude of whose superficial volume is also denoted by the same symbol τ_0).

After introduction of an impulsive source $\mathbf{U}\,\delta(\mathbf{R} - \mathbf{R}')\,\delta(t)$ of momentum into the suspension at position $\mathbf{R} = \mathbf{R}'$ (with $\mathbf{R}' = \mathbf{R}_{n'} + \mathbf{r}'$), at time $t = 0$ and vector strength **U**, subsequent transport of this momentum tracer through the suspension (the latter assumed to be initially at rest) is governed by the system of equations (Mauri and Brenner, 1991a,b)

$$\partial \mathbf{P}/\partial t + \mathbf{\nabla} \cdot \ddot{\mathbf{J}} = \mathbf{I}\,\delta_{m0}\,\delta(\mathbf{r} - \mathbf{r}')\,\delta(t), \tag{8.10}$$

$$\mathbf{\nabla} \cdot \mathbf{P} = 0, \tag{8.11}$$

with

$$\ddot{\mathbf{J}} = \mathbf{I}\Pi - \nu(\mathbf{r})(\mathbf{\nabla}\mathbf{P} + {}^\dagger\mathbf{\nabla}\mathbf{P}). \tag{8.12}$$

Here, **I** is the dyadic idemfactor and $\mathbf{m} = \mathbf{n} - \mathbf{n}'$; moreover, $\mathbf{\nabla} \equiv \partial/\partial \mathbf{r}$ is the local gradient operator defined only within a unit cell ($\mathbf{r} \in \tau_0$). The "material" functions **P**, $\ddot{\mathbf{J}}$, and Π are the respective dyadic, triadic, and vector fields defined by the linear transformations

$$\mathbf{p} = \mathbf{P}(\mathbf{m}, \mathbf{r}, t\,|\,\mathbf{r}') \cdot \mathbf{U} \qquad (p_i = P_{ij}U_j), \tag{8.13a}$$

$$\pi = \Pi(\mathbf{m}, \mathbf{r}, t\,|\,\mathbf{r}') \cdot \mathbf{U} \qquad (\pi = \Pi_j U_j), \tag{8.13b}$$

$$\mathbf{T} = -\ddot{\mathbf{J}}(\mathbf{m}, \mathbf{r}, t\,|\,\mathbf{r}') \cdot \mathbf{U} \qquad (T_{ij} = -J_{ijk}U_k), \tag{8.13c}$$

as the respective "propagators" for the momentum density $\mathbf{p} \equiv \rho\mathbf{v}$, the pressure π, and the total stress tensor **T**, each of which arise from the introduction of the impulse of strength **U** at position $\mathbf{R}'$ at time $t = 0$. That these are indeed material functions follows from the fact that these fields depend only on the prescribed kinematic viscosity field $\nu(\mathbf{r})$ and spatially periodic configuration—which functional dependence follows from the uniquely posed system of Eqs. (8.10)–(8.13) and (8.14) that define these fields.

Equations (8.10)–(8.13) are merely Stokes equations rewritten in a suggestive form chosen to emphasize transport of the momentum tracer density $\rho\mathbf{v}$, as well as to exploit the analogy between Eq. (8.10) for momentum transport and the comparable equation (Brenner, 1980b) for transport of the scalar probability density P, which is equivalent to the material tracer density. The absence of a convective term **vp** from the flux expression in Eq. (8.12) reflects the neglect of inertial effects (valid at small, unit-cell Reynolds numbers), where transport occurs purely by the "molecular diffusion" of momentum. Explicitly, the dyadic $P_{ij}(\mathbf{R}, t | \mathbf{R}')$ [note the equivalence of the arguments $(\mathbf{R}, t | \mathbf{R}')$ and $(\mathbf{m}, \mathbf{r}, t | \mathbf{r}')$] may be regarded as the dyadic "probability density" (more conventionally, the Green's function) for finding the momentum tracer at position $\mathbf{R}$ at time t pointing in the x_i direction ($i = 1, 2, 3$), given that initially ($t = 0$) it was located at $\mathbf{R}'$ and had direction x_j ($j = 1, 2, 3$). The triadic $\ddot{\mathbf{J}}$ may be similarly interpreted as the flux of the momental probability density arising in joint consequence of momental "Brownian" motion (whose intensity is measured by the kinematic viscosity) together with the action of the pressure stresses Π [cf. Eq. (8.12)].

Equations (8.10)–(8.12) are to be solved within each unit cell subject to the conditions

$$(\mathbf{P}, \hat{\mathbf{n}} \cdot \ddot{\mathbf{J}}) \quad \text{continuous across } S_p, \tag{8.14a}$$

$$|\mathbf{R}_m|^n \mathbf{h} \to 0 \text{ as } |\mathbf{m}| \to \infty \quad (n = 0, 1, 2, \ldots), \tag{8.14b}$$

and

$$\mathbf{h}(\mathbf{m}^*, \mathbf{r} + \mathbf{l}_j) = \mathbf{h}(\mathbf{m}^*, \mathbf{r} + \mathbf{I}_j) = \mathbf{h}(\mathbf{m}, \mathbf{r}) \quad \text{on } \partial \tau_0. \tag{8.15}$$

Here, $\hat{\mathbf{n}}$ is a unit normal vector at the interstitial phase boundary S_p, if any, across which the kinematic viscosity $v(\mathbf{r})$ is possibly discontinuous, whereas the generic symbol $\mathbf{h}$ denotes either $\mathbf{P}$ or $\ddot{\mathbf{J}}$. Equation (8.15) expresses continuity of the pertinent fields across the unit cell faces $\partial \tau_0$. In Eq. (8.15), $\mathbf{m}^*$ is to be chosen such that $\mathbf{R}_{\mathbf{m}^*} = \mathbf{R}_\mathbf{m} - \mathbf{I}_j$ is a basic lattice vector [cf. Eq. (7.1)].

Equations (8.10)–(8.12), tensorial ranks and boundary conditions (8.14)–(8.15) notwithstanding, embody a structure similar in format and symbolism to their counterparts for the transport of passive scalars, e.g., the material transport of the scalar probability density P (Brenner, 1980b; Brenner and Adler, 1982), at least in the absence of convective transport. As such, by analogy to the case of nonconvective material transport, the effective kinematic viscosity $\bar{v}_{ijkl}$ of the suspension may be obtained by matching the total spatial moments of the probability density P_{ij} to those of an equivalent coarse-grained dyadic probability density $\bar{P}_{ij}$, valid on the suspension scale, using a scheme (Brenner and Adler, 1982) identical in conception to that used to determine the effective diffusivity for material transport at the Darcy scale from the analogous scalar material probability density P. In particular, the second-order total moment $\ddddot{\mathbf{M}}^{(2)}$ ($\equiv \mathbf{M}^{(2)}_{ijkl}$) of the probability density $\mathbf{P}$, defined as

$$\ddddot{\mathbf{M}}^{(2)}(t | \mathbf{r}') = \int_{\tau_0} d\mathbf{r} \sum_\mathbf{m} \mathbf{P}(\mathbf{m}, \mathbf{r}, t | \mathbf{r}') \mathbf{R}_\mathbf{m} \mathbf{R}_\mathbf{m}, \tag{8.16}$$

is found to be governed in the asymptotic long-time limit by the expression

$$d\dddot{\mathbf{M}}^{(2)}/dt \simeq \tau_0^{-1} \int_{\partial \tau_0} d\mathbf{s} \cdot [\dddot{\mathcal{J}}\mathbf{r} + (\dddot{\mathcal{J}}\mathbf{r})^\dagger]. \tag{8.17}$$

In Eq. (8.17), $\dddot{\mathcal{J}} (\equiv \mathcal{J}_{ijkl})$ is found by solving the steady-state unit cell equations (Mauri and Brenner, 1991a)

$$\mathbf{V} \cdot \dddot{\mathcal{J}} = \mathbf{0}, \tag{8.18a}$$

$$\mathbf{V} \cdot \dddot{\mathcal{P}} = \mathbf{0}, \tag{8.18b}$$

$$\dddot{\mathcal{J}} = \mathbf{I}\dddot{p} - v(\nabla \dddot{\mathcal{P}} + {}^\dagger\nabla \dddot{\mathcal{P}}), \tag{8.18c}$$

$$(\dddot{\mathcal{P}}, \hat{\mathbf{n}} \cdot \dddot{\mathcal{J}}) \text{ continuous across } S_p. \tag{8.18d}$$

The three tensor fields $\dddot{p}(\equiv p_{ij})$, $\dddot{\mathcal{P}} \equiv (\mathcal{P}_{ijk})$ and $\dddot{\mathcal{J}}(\equiv \mathcal{J}_{ijkl})$ appearing in these equations are further required to satisfy the "jump" boundary conditions

$$[\![\dddot{\mathcal{P}}]\!] = -[\![\mathbf{Ir}]\!], \tag{8.19a}$$

$$[\![\dddot{p}]\!] = \mathbf{0}, \tag{8.19b}$$

$$[\![\dddot{\mathcal{J}}]\!] = \mathbf{0}, \tag{8.19c}$$

imposed across the cell boundaries $\partial \tau_0$. The double-bracketed operator $[\![\ ]\!]$ appearing in Eq. (8.19) denotes the jump in the value of its argument across "equivalent" points symmetrically situated on opposite faces of the unit cell.

The suspension-scale momentum transport equation, whose moments need to be matched with those of Eq. (8.17), is adopted (Mauri and Brenner, 1991a) in the general form

$$\partial \bar{\mathbf{P}}/\partial t + \mathbf{V}_\mathbf{R} \cdot \dddot{\mathbf{J}} = \mathbf{I} \delta(\mathbf{R} - \mathbf{R}')\delta(t) \tag{8.20a}$$

$$\mathbf{V}_\mathbf{R} \cdot \bar{\mathbf{P}} = \mathbf{0}, \tag{8.20b}$$

with

$$\dddot{\mathbf{J}} = \mathbf{I}\bar{\Pi} - 2\ddddot{v}:\mathbf{V}_\mathbf{R}\bar{\mathbf{P}}. \tag{8.20c}$$

An overbar denotes coarse-grained quantities. The Newtonian constitutive Eq. (8.20c), in effect, defines the configuration-specific, anisotropic, kinematic viscosity-tetradic $\ddddot{v}$ ($=\bar{v}_{ijkl}$) of the spatially periodic suspension. Subject to the attenuation conditions,

$$|\mathbf{R} - \mathbf{R}'|^n(\bar{\mathbf{P}}, \dddot{\mathbf{J}}) \to 0 \quad \text{as } |\mathbf{R} - \mathbf{R}'| \to \infty \quad (n = 0, 1, 2, \ldots), \tag{8.20d}$$

the set of Eq. (8.20) is used to derive evolution equations for the total moments of the suspension-scale probability dyadic $\bar{\mathbf{P}}$. The second-order moment of $\bar{\mathbf{P}}$

is found to adopt the asymptotic form

$$d\overset{....}{\mathbf{M}}{}^{(2)}/dt \simeq 4\overset{....}{\bar{v}}. \qquad (8.21)$$

Mauri and Brenner (1991a) show the lower-order moments to be properly matched; furthermore, the asymptotic correspondence of Eq. (8.21) with Eq. (8.17) requires that

$$\overset{....}{\bar{v}}_{ijkl} = \tfrac{1}{2}(N_{ijkl} + N_{ljki}), \qquad (8.22)$$

where

$$\overset{....}{\mathbf{N}} = \frac{1}{2\tau_0}\int_{\partial\tau_0} \mathbf{r}\, ds \cdot \overset{....}{\mathscr{J}} \qquad (8.23)$$

with $\overset{....}{\mathscr{J}}$ to be found by solving Eq. (8.18) and (8.19).

Equation (8.22) constitutes the means whereby the configuration-specific kinematic viscosity of the suspension may be computed from the prescribed spatially periodic, microscale, kinematic viscosity data $v(\mathbf{r})$ by first solving an appropriate microscale unit-cell problem. Its "Lagrangian" derivation differs significantly from volume-average Eulerian approaches (Zuzovsky et al., 1983; Nunan and Keller, 1984) usually employed in deriving such suspension-scale properties.

Results for the rigid-particle case can be extracted (Mauri and Brenner, 1991a) from the analysis presented here by formally setting

$$v(\mathbf{r}) = v_0 \equiv \text{const} \qquad (\mathbf{r} \in \tau_f) \qquad (8.24)$$

in Eq. (8.18c) and solving Eqs. (8.18) and (8.19) in the interstitial fluid region τ_f external to the rigid-particle surfaces S_p so as to satisfy the "no-slip" boundary condition

$$\mathbf{P} = \rho(\mathbf{V} + \mathbf{\Omega} \times \mathbf{r}) \quad \text{on } S_p. \qquad (8.25)$$

Here, the vector constants $\mathbf{V}$ and $\mathbf{\Omega}$ are to be determined so as to satisfy the respective pair of dyadic and triadic particle–surface boundary conditions

$$\int_{S_p} d\mathbf{s} \cdot \overset{....}{\mathscr{J}} = \mathbf{0}, \qquad (8.26a)$$

$$\int_{S_p} \mathbf{r} \times (d\mathbf{s} \cdot \overset{....}{\mathscr{J}}) = \mathbf{0}, \qquad (8.26b)$$

which express the force- and torque-free properties of the suspended bodies (and hence of the suspension as a whole).

To recapitulate, the momental "Brownian" motion of the conserved momentum tracer allows it to sample all portions of the suspension's intracell kinematic viscosity field $v(\mathbf{r})$, $(\mathbf{r} \in \tau_0)$, thereby furnishing the proper effective viscosity of the suspension *without* the necessity for defining any arbitary averages of the microscale stresses or velocity gradients. Such ad hoc averages have heretofore constituted an essential element of virtually all suspension rheology schemes. Appearing in its stead is an Einstein-type diffusion formula for the temporal spread of the second moment of the momentum density, which provides the basis for the scheme. Equally novel is that the suspension is macroscopically at rest during the momentum diffusion process, rather than undergoing steady shear. This essentially Lagrangian approach furnishes the configuration-specific kinematic viscosity of concentrated suspensions. The configuration-specific values obtained by this scheme can be shown (Mauri and Brenner, 1991b) to be identical to values derived from the scheme described in Section VII.

Falling-Ball Suspension Viscometry

Closely related to the preceding momentum tracer scheme, which utilizes (tracer) motion through an otherwise quiescent suspension to establish the suspension's rheological properties, is the experimental work of Mondy, Graham and co-workers (1986a,b, 1987). Regarding each of their suspensions of neutrally buoyant particles as being a fluid continuum possessing an effective Newtonian scalar viscosity μ, they used experimental observations of the mean (i.e., time average) sedimentation speed of a heavy ball falling through the suspension, together with an assumed applicability of Stokes law to this mean ball velocity, to determine μ (after extrapolating their results to eliminate wall effects). Results obtained in this manner agreed well, over the entire concentration range studied ($0 < \phi < 0.5$), with standard nonquiescent suspension viscosity data (Thomas, 1965; Gadala-Maria and Acrivos, 1980) derived from Couette and capillary flow experiments, during which, the suspension was, of course, sheared.

What is remarkable about the Mondy-Graham results is that for each suspension investigated, the observed viscosity was independent of the relative diameter of the sedimenting ball (d_f) to that of the suspended spheres (d_s) over the entire size range studied ($0.75 \leq d_f/d_s \leq 12.0$). This despite the fact that, because of collisions with the suspended spheres, the instantaneous trajectory of the falling ball was observed to deviate significantly from a simple, uniform-speed, purely vertical trajectory at the smaller ball sizes. In the words of Mondy *et al.*, (1986a),

> Passage of a ball of the same or smaller diameter than the suspended balls is extremely discontinuous. Periods of almost no motion, as the falling ball approaches and 'rolls off' suspended spheres, alternate with periods of almost free fall in the interstices between suspended particles. A statistical analysis reveals that falling-ball terminal velocities averaged over a distance of approximately 100 suspended-ball diameters are reproducible.

It appears to us that this apparently "random" motion, superposed on the steady mean-settling velocity, is analogous to that of Brownian motion in the sense that it arises from the fact that the continuum is not structureless.

Subsequent falling-ball experiments performed with suspended rods (Graham et al., 1987; Milliken et al., 1989) replacing the spheres revealed significant and systematic differences between quiescent values of the suspension viscosity and those derived from Couette and capillary viscometer flow measurements. This is attributed by Graham, Mondy, and co-workers to fundamental differences in the distributions of rod orientations characterizing the quiescent and sheared suspensions.

C. Fractal Suspensions

Another contemporary research thrust described here pertains to the hydrodynamic properties of fractal suspensions. The potential importance of such structures stems from the close analogy existing between percolation and suspension rheology near the critical concentration, as outlined in Section VI. Since the infinite cluster formed near the percolation threshold is fractal, the suspension itself may be expected to be fractal in that limit. A second potential area of application is to the rheology of highly polydisperse suspensions, involving a multiplicity of length scales and hence, suggesting the existence of fractal structures. Such structures may be expected to arise here from the presence of the large variety of particle sizes rather than from formation of infinite clusters, which was previously the case in the percolation analog.

Aggregable suspensions were shown (Weitz and Oliveria, 1984; Schaefer et al., 1984) to yield fractal flocs, a result which sheds new light on the old subject of coagulation and flocculation.

Available results pertinent to the hydrodynamics of fractal suspensions are sparse thus far, encompassing only three physical situations. Gilbert and Adler (1986) determined the Stokes rotation-resistance dyadic for spheres arranged in a Leibniz packing [Fig. 7(a)]. With the gap between any two spheres assumed small compared with their radii, lubrication-type approximations suffice. In this analysis, the inner spheres are assumed to rotate freely, whereas external torques $\mathbf{T}_i$ ($i = 1, 2, 3$) are applied to the three other spheres. For Stokes flow, these torques are linearly related to the sphere angular velocities by the expression

$$(\mathbf{T}_1, \mathbf{T}_2, \mathbf{T}_3) = \mathbf{M}_n \cdot (\omega_1, \omega_2, \omega_3), \tag{8.27}$$

with $\mathbf{M}_n$ as the rotational resistance dyadic whose numerical value is dependent on the generation number n of the packing. Approximate formulas exist for $\mathbf{M}_n$; moreover, its asymptotic behavior for large n has been studied.

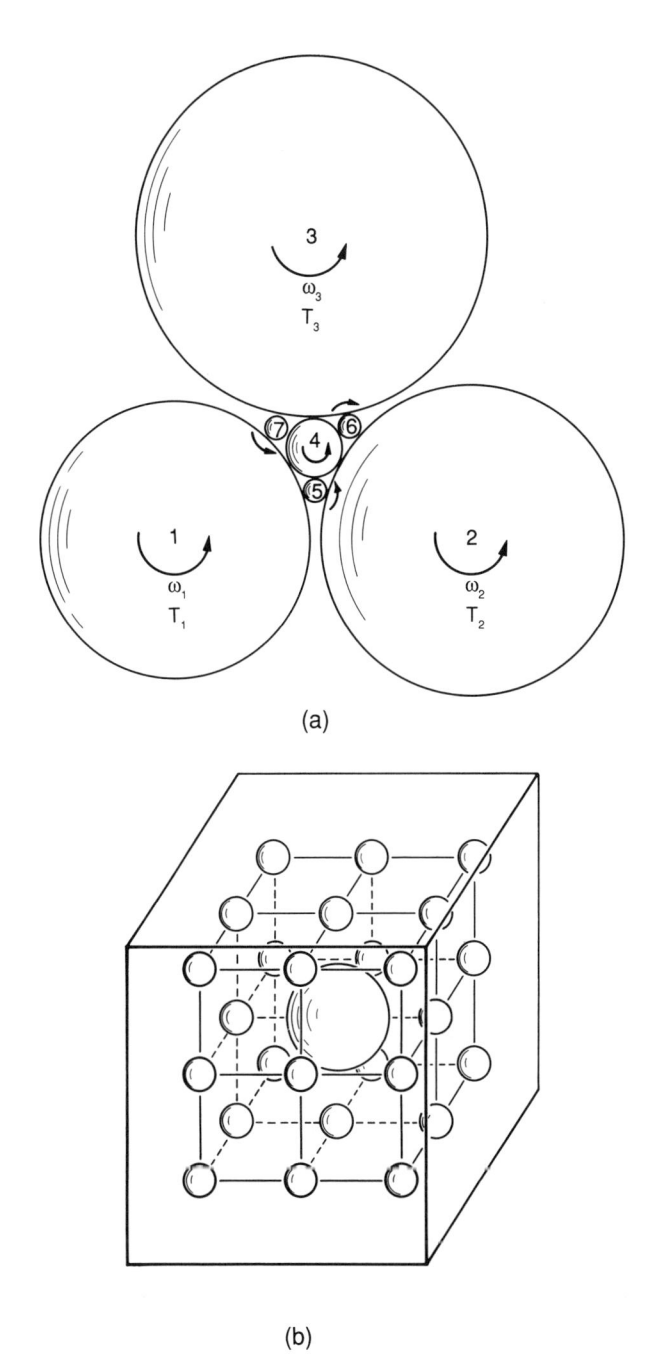

FIG. 7. Two examples of fractal suspensions (a) The spheres are arranged in a Leibnitz packing with the construction process illustrated to $n = 2$; sphere 4 is created during the generation $n = 1$, while spheres 5, 6 and 7 are created during step $n = 2$. (b) The spheres are arranged according to a modified Menger sponge, again the contruction stage is shown to $n = 2$.

Another fractal structure of interest is considered by Adler (1986). A three-dimensional fractal suspension may be constructed from a modified Menger sponge, as shown in Fig. 7(b). A scaling argument permitted calculating the effective viscosity of such a suspension; however, this viscosity should be compared with numerical results for the solution of Stokes equations in such a geometry before this rheological result is accepted unequivocally.

Finally, in a related dynamical, though nonrheological context, Adler (1987) determined the drag exerted on fractal two-dimensional flocs possessing various fractal dimensions. He found a very weak effect (if indeed any effect) of this particular parameter on the drag for spatial dimensions ranging from 1.5 to 2.0.

D. Antisymmetric Stresses, Internal Spin Fields, and Vortex Viscosity in Magnetic Fluids

Suspended spherical particles, each containing a permanently embedded dipole (e.g., magnetic), are unable to freely rotate (Brenner, 1984; Sellers and Brenner, 1989) in response to the shear and/or vorticity field that they are subjected to whenever a complementary external (e.g., magnetic) field acts on them. This hindered rotation results from the tendency of the dipole to align itself parallel to the external field because of the creation of a couple arising from any orientational misalignment between the directions of the dipole and external field. In accordance with Cauchy's moment-of-momentum equation for continua, these couples in turn give rise to an antisymmetric state of stress in the dipolar suspension, representable as the pseudovector $\mathbf{T}_\times = -\overset{\leftrightarrow}{\varepsilon}:\mathbf{T}^a$ of the antisymmetric portion $\mathbf{T}^a = \frac{1}{2}(\mathbf{T} - \mathbf{T}^\dagger)$ of the deviatoric stress $\mathbf{T} = \mathbf{P} + \mathbf{I}p$.

Accompanying the impeded particle rotation is the (kinematical) existence of an internal spin field $\mathbf{\Omega}$ within the suspension, which is different from one-half the vorticity $\omega = (\frac{1}{2})\nabla \times \mathbf{v}$ of the suspension. The disparity $\omega - \mathbf{\Omega}$ between the latter two fields serves as a reference-frame invariant pseudovector in the constitutive relation $\mathbf{T}_\times = \zeta(\omega - \mathbf{\Omega})$, which defines the so-called "vortex viscosity" ζ of the suspension. Expressions for $\zeta(\phi)$ as a function of the volume ϕ of suspended spheres are available (Brenner, 1984) over the entire particle concentration range and are derived from the prior calculations of Zuzovsky et al. (1983) for cubic, spatially-periodic suspension models.

These suspension viscosity concepts are of growing technological importance in rationalizing and quantifying the behavior and properties of so-called magnetic fluids (Rosensweig, 1982, 1985, 1987). In a novel proposal, Brenner (1984) outlined a potentially useful scheme to use the apparently rigid-body rotation of a dipolar suspension to measure its vortex viscosity (independently of its shear viscosity). Closely related to vortex viscosity is the heat- and mass-transfer study by Nadim *et al.* (1986) of enhanced, effective-conductivity transport properties in such dipolar suspensions when undergoing apparently rigid-body rotations. The so-called continuum-mechanical principle of material-frame indifference (Ryskin and Rallison, 1980) would, in such circumstances, incorrectly suggest the impossibility of affecting the transport rate by a rigid-body rotation of the apparatus housing the suspension.

Intimately related to these magnetic-field suspension rheology developments is the growing field of electrorheology, which involves comparable electric fields and was the subject of an international symposium (Carlson and Conrad, 1987).

ACKNOWLEDGEMENT

Writing of this chapter was facilitated by grants to Howard Brenner from the Office of Basic Energy Sciences of the Department of Energy and the National Science Foundation. The authors also wish to thank the referees, Professors Andreas Acrivos and William B. Russel, for bringing a number of relevant papers to our attention.

References

Ackerson, B. J., and Clark, N. A., *Physica A (Amsterdam)* **118A,** 221 (1983).
Adler, P. M., *AIChE J.* **25,** 487 (1979).
Adler, P. M., *J. Colloid Interface Sci.* **84,** 461 (1981a).
Adler, P. M., *J. Colloid Interface Sci.* **84,** 475 (1981b).
Adler, P. M., *J. Mec. Theor. Appl.* **3,** 725 (1984).
Adler, P. M., *Phys. Fluids* **29,** 15 (1986).
Adler, P. M., *Faraday Discuss. Chem. Soc.* **83,** 145 (1987).
Adler, P. M., and Brenner, H., *Int. J. Multiphase Flow* **11,** 361 (1985a).
Adler, P. M., and Brenner, H., *J. Phys., Colloq. (Orsay, Fr.)* **46** (C-3), 223 (1985b).
Adler, P. M., and Mills, P. M., *J. Rheol.* **23,** 25 (1979).
Adler, P. M., Zuzovsky, M., and Brenner, H., *Int. J. Multiphase Flow* **11,** 387 (1985).
Ansell, G. C., and Dickinson, E., *J. Colloid Interface Sci.* **110,** 73 (1986).
Aref, H., and Balachandar, S., *Phys. Fluids* **29,** 3515 (1986).
Arp, P. A., and Mason, S. G., *Colloid Polym. Sci.* **255,** 566 (1977a).
Arp, P. A., and Mason, S. G., *J. Colloid Interface Sci.* **61,** 21 (1977b).
Aubert, J. H., Kraynik, A. M., and Rand, P. B., *Sci. Am.* **254,** 74 (1986)
Barnes, H. A., Edwards, M. F., and Woodcock, L. V., *Chem. Eng. Sci.* **42,** 591 (1987).
Batchelor, G. K., *J. Fluid Mech.* **41,** 545 (1970).
Batchelor, G. K., *J. Fluid Mech.* **52,** 245 (1972).
Batchelor, G. K., *Annu. Rev. Fluid Mech.* **6,** 227 (1974).
Batchelor, G. K., *Proc. Int. Union Theor. Appl. Mech., Delft Congr.* (W. T. Koiter, ed.)., pp. 33–55. North-Holland Publ., Amsterdam, 1976a.
Batchelor, G. K., *J. Fluid Mech.* **74,** 1 (1976b).

Batchelor, G. K., *J. Fluid Mech.* **83**, 97 (1977).
Batchelor, G. K., *J. Fluid Mech.* **119**, 379 (1982); Corrigendum, *J. Fluid Mech.* **137**, 467 (1983).
Batchelor, G. K., *J. Fluid Mech.* **131**, 155 (1983); Corrigendum, *J. Fluid Mech.* **137**, 467 (1983).
Batchelor, G. K., and Green, J. T., *J. Fluid Mech.* **56**, 375 (1972a).
Batchelor, G. K., and Green, J. T., *J. Fluid Mech.* **56**, 401 (1972b).
Batchelor, G. K., and Wen, C.-S., *J. Fluid Mech.* **124**, 495 (1982); Corrigendum, *J. Fluid Mech.* **137**, 467 (1983).
Bedeaux, D., Kapral, R., and Mazur, P., *Physica A (Amsterdam)* **88A**, 88 (1977).
Beenakker, C. W. J., *Physica A (Amsterdam)* **128A**, 48 (1984).
Beenakker, C. W. J., and Mazur, P., *Phys. Lett.* **91**, 290 (1982).
Beenakker, C. W. J., and Mazur, P., *Physica A (Amsterdam)* **120A**, 388 (1983).
Beenakker, C. W. J., and Mazur, P., *Physica A (Amsterdam)* **126A**, 349 (1984).
Bensoussan, A., Lions, J. L., and Papanicolaou, G., "Asymptotic Analysis for Periodic Structures." North-Holland Publ., Amsterdam, 1978.
Bentley, B. J., and Leal, L. G., *J. Fluid Mech.* **167**, 219 (1986a).
Bentley, B. J., and Leal, L. G., *J. Fluid Mech.* **167**, 241 (1986b).
Berryman, J. G., *Phys. Rev.* **A27**, 1053 (1983).
Bird, R. B., Curtiss, C. F., Armstrong, R. C., and Hassager, O., "Dynamics of Polymeric Liquids. Vol. 2: Kinetic Theory," 2nd Ed. Wiley (Interscience), New York, 1987.
Bossis, G., and Brady, J. F., *J. Chem. Phys.* **80**, 5141 (1984).
Bouillot, J. L., Camoin, C., Belzons, M., Blanc, R., and Guyon, E., *Adv. Colloid Interface Sci.* **17**, 299 (1982).
Brady, J. F., and Bossis, G., *J. Fluid Mech.* **155**, 105 (1985).
Brady, J. F., and Bossis, G., *Annu. Rev. Fluid Mech.* **20**, 111 (1988).
Brenner, H., *Phys. Fluids* **1**, 338 (1958).
Brenner, H., *Annu. Rev. Fluid Mech.* **2**, 137 (1970).
Brenner, H., *Prog. Heat Mass Transfer* **5**, 89 (1972a).
Brenner, H., *Prog. Heat Mass Transfer* **6**, 509 (1972b).
Brenner, H., *Int. J. Multiphase Flow* **1**, 195 (1974).
Brenner, H., *PhysicoChem. Hydrodyn.* **1**, 91 (1980a).
Brenner, H., *Philos. Trans. R. Soc. London, Ser. A* **297**, 81 (1980b).
Brenner, H., *PhysicoChem. Hydrodyn.* **3**, 139 (1982).
Brenner, H., *Int. J. Eng. Sci.* **22**, 645 (1984).
Brenner, H., and Adler, P. M., *Philos. Trans. R. Soc. London, Ser. A* **307**, 149 (1982).
Brenner, H., and O'Neill, M. E., *Chem. Eng. Sci.* **27**, 1421 (1972).
Brinkman, H. C., *Appl. Sci. Res., Sect. A* **1**, 27 (1947).
Broadbent, S. R., and Hammersley, J. M., *Proc. Cambridge Philos. Soc.* **53**, 629 (1957).
Brunn, P., *J. Non-Newtonian Fluid Mech.* **7**, 271 (1980).
Brunn, P., *Int. J. Multiphase Flow* **7**, 221 (1981).
Buscall, R., Goodwin, J. W., Ottewill, R. H., and Tadros, T. F., *J. Colloid Interface Sci.* **85**, 78 (1982).
Buyevich, Y. A., and Shchelchkova, I. N., *Prog. Aerosp. Sci.* **18**, 121 (1978).
Caflish, R. E., and Luke, J. H. C., *Phys. Fluids* **28**, 759 (1985).
Camoin, C., and Blanc, R., *J. Phys. Lett.* **46**, L67 (1985).
Carlson, J. D., and Conrad, H., organizers, *Int. Symp. Electrorheol. Fluids, Annu. Meet. Fine Part. Soc. 18th, Boston Mass.* (1987).
Castillo, C. A., Rajagopalan, R., and Hirtzel, C. S., *Rev. Chem. Eng.* **2**, 237 (1984).
Childress, S. *J. Chem. Phys.* **56**, 2527 (1972).
Cichocki, B., Felderhof, B. U., and Schmitz, R., *PhysicoChem. Hydrodyn.* **10**, 383 (1988).
Collet, P., and Eckmann, J.-P., "Iterated Maps on the Interval as Dynamic Systems." Birkhaeuser, Boston, Massachusetts, 1980.

Cox, R. G., and Brenner, H., *Chem. Eng. Sci.* **26**, 65 (1971).
Cox, R. G., and Mason, S. G., *Annu. Rev. Fluid Mech.* **3**, 291 (1971).
Darabaner, C. L., and Mason, S. G., *Rheol. Acta* **6**, 273 (1967).
Davis, A. M. J., and Brenner, H., *J. Eng. Mech. Div., Am. Soc. Civ. Eng.* **107**, 609 (1981).
Davis, R. H., and Acrivos, A., *Annu. Rev. Fluid Mech.* **17**, 91 (1985).
DeGennes, P. G., *J. Phys. (Orsay, Fr.)* **40**, 783 (1979).
DeGennes, P. G., *PhysicoChem. Hydrodyn.* **2**, 31 (1981).
de Kruif, C. G., van Iersel, E. M. F., Vrij, A., and Russel, W. B., *J. Chem. Phys.* **83**, 4717 (1985).
Derrida, B., and Vannimenus, J., *J. Phys. A* **15**, L557 (1982).
Durlofsky, L., Brady, J. F., and Bossis, G., *J. Fluid Mech.* **180**, 21 (1987).
Eilers, H., *Kolloid-Z.* **97**, 313 (1941).
Einstein, A., *Ann. Phys.* **19**, 289 (1906).
Einstein, A., *Ann Phys.* **34**, 591 (1911).
Einstein, A., in "The Theory of the Brownian Movement" (R. Fürth, ed.). Dover, New York, 1956.
Evans, D. J., *Mol. Phys.* **37**, 1745 (1979).
Felderhof, B. U., *Physica A (Amsterdam)* **89A**, 373 (1977).
Felderhof, B. U., *J. Phys. A: Math Gen.* **11**, 929 (1978).
Felderhof, B. U., *Physica A (Amsterdam)* **147A**, 533 (1988).
Feuillebois, F., *J. Fluid Mech.* **139**, 145 (1984).
Feuillebois, F., in "Multiphase Science and Technology" (G. S. Hewitt, J.-M. Delhaye, and N. Zuber, eds.). Hemisphere, New York, 1988.
Fitch, E. B., *AIChE J.* **25**, 913 (1979).
Frankel, N. A., and Acrivos, A., *Chem. Eng. Sci.* **22**, 847 (1967).
Freed, K. F., and Muthukumar, M., *J. Chem. Phys.* **68**, 2088 (1978).
Freed, K. F., and Muthukumar, M., *J. Chem. Phys.* **76**, 6186 (1982).
Gadala-Maria, F., and Acrivos, A., *J. Rheol.* **24**, 799 (1980).
Ganatos, P., Pfeffer, R., and Weinbaum, S., *J. Fluid Mech.* **84**, 79 (1978).
Gilbert, F. J., and Adler, P. M., *J. Colloid Interface Sci.* **114**, 243 (1986).
Glendinning, A. B., and Russel, W. B., *J. Colloid Interface Sci.* **89**, 124 (1982).
Goldman, A. J., Cox, R. G., and Brenner, H., *Chem. Eng. Sci.* **21**, 1151 (1966).
Goldman, A. J., Cox, R. G., and Brenner, H., *Chem. Eng. Sci.* **22**, 637 (1967a).
Goldman, A. J., Cox. R. G., and Brenner, H., *Chem. Eng. Sci.* **22**, 653 (1967b).
Goldsmith, H. L., and Mason, S. G., in "Rheology: Theory and Applications" (F. R. Eirich, ed.), Vol. 4, pp. 85–250. Academic Press, New York, 1967.
Goto, H., and Kuno, H., *J. Rheol.* **26**, 387 (1982).
Graham, A. L., Mondy, L. A., Gottlieb, M., and Powell, R. L., *Appl. Phys. Lett.* **50**, 127 (1987).
Guth, E. von, and Simha, R., *Kolloid-Z.* **74**, 266 (1936).
Haber, S., and Brenner, H., *J. Colloid Interface Sci.* **97**, 496 (1984).
Haber, S., and Hetsroni, G., *J. Colloid Interface Sci.* **79**, 56 (1981).
Hansen, J. P., and McDonald, I. R., "Theory of Simple Liquids." Academic Press, New York, 1976.
Happel, J., *J. Appl. Phys.* **28**, 1288 (1957).
Happel, J., *AIChE J.* **4**, 197 (1958).
Happel, J., and Brenner, H., "Low Reynolds Number Hydrodynamics." Prentice-Hall, Englewood Cliffs, New Jersey, 1965.
Hassonjee, Q., Ganatos, P., and Pfeffer, R., *J. Fluid Mech.* **197**, 1 (1988).
Havlin, S., and Ben-Arraham, D., *J. Phys. A: Math. Gen.* **16**, L483 (1983).
Herczyński, R., and Pieńkowska, I., *Annu. Rev. Fluid Mech.* **12**, 237 (1980).
Hinch, E. J., *J. Fluid Mech.* **54**, 423 (1972).
Hinch, E. J., *J. Fluid Mech.* **83**, 695 (1977).

Howells, I. D., *J. Fluid Mech.* **64,** 449 (1974).
Husband, D. M., and Gadala-Maria, F., *J. Rheol.* **31,** 95 (1987).
Itoh, S., *J. Phys. Soc. J.* **52,** 2379 (1983).
Jeffrey, D. J., *Proc. R. Soc. London Ser., A* **338,** 503 (1974).
Jeffrey, D. J., and Acrivos, A., *AIChE. J.* **22,** 417 (1976).
Jeffrey, D. J., and Onishi, Y., *J. Fluid Mech.* **139,** 261 (1984).
Jinescu, V. V., *Int. Chem. Eng.* **14,** 397 (1974).
Kao, S. V., Cox, R. G., and Mason, S. G., *Chem. Eng. Sci.* **32,** 1505 (1977).
Kapral, R., and Bedeaux, D., *Physica A (Amsterdam)* **91A,** 590 (1978).
Karnis, A., Goldsmith, H. L., and Mason, S. G., *J. Colloid Interface Sci.* **22,** 531 (1966).
Kim, S., *Int. J. Multiphase Flow* **12,** 469 (1986).
Kim, S., and Mifflin, R. T., *Phys. Fluids* **28,** 2033 (1985).
Kops-Werkhoven, M. M., and Fijnaut, H. M., *J. Chem. Phys.* **74,** 1618 (1981).
Kops-Werkhoven, M. M., and Fijnaut, H. M., *J. Chem. Phys.* **77,** 2242 (1982).
Kraynik, A. M., *Annu. Rev. Fluid Mech.* 20, 325 (1988).
Kraynik, A. M., and Hansen, M. G., *J. Rheol.* **30,** 409 (1986).
Kraynik, A. M., and Hansen, M. G., *J. Rheol.* **31,** 175 (1987).
Krieger, I. M., and Dougherty, T. J., *Trans. Soc. Rheol.* **3,** 137 (1959).
Kynch, G. J., *J. Fluid Mech.* **5,** 193 (1959).
Lamb, M., "Hydrodynamics," 6th Ed. Cambridge Univ. Press, Cambridge, England, 1932.
Landau, L. D., and Lifschitz, E. M., "Fluid Mechanics." Pergamon, London, 1959.
Leal, L. G., *J. Non-Newtonian Fluid Mech.* **5,** 33 (1979).
Leal, L. G., *Annu. Rev. Fluid Mech.* **12,** 435 (1980).
Leichtberg, S., Weinbaum, S., Pfeffer, R., and Gluckman, M. J., *Philos. Trans. R. Soc. London,* **282,** 585 (1976).
Leighton, D., and Acrivos, A., *J. Fluid Mech.* **181,** 415 (1987).
Lekkerkerker, C. G., "Geometry of Numbers." North-Holland Publ., Amsterdam, 1969.
Lévy, T., and Sanchez-Palencia, E., *C. R. Acad. Sci., Ser. 2* **297,** 193 (1983a).
Lévy, T., and Sanchez-Palencia, E., *J. Non-Newtonian Fluid Mech.* **13,** 63 (1983b).
Lundgren, T. S., *J. Fluid Mech.* **51,** 273 (1972).
McQuarrie, D. A., "Statistical Mechanics." Harper, New York, 1976.
Manley, R. St. J., and Mason, S. G., *J. Colloid Sci.* **7,** 354 (1952).
Manley, R. St. J., and Mason, S. G., *Can. J. Chem.* **32,** 763 (1954).
Marrucci, G., and Denn, M. M., *Rheol Acta* **24,** 317 (1985).
Mason, S. G., *J. Colloid Interface Sci.* **58,** 275 (1977).
Maude, A. D., and Whitmore, R. L., *J. Appl. Phys.* **9,** 477 (1958).
Mauri, R., and Brenner, H., to be published (1991a).
Mauri, R., and Brenner, H., to be published (1991b).
Mazur, P., *Physica A (Amsterdam)* **110A,** 128 (1982).
Mazur, P., *Faraday Discuss. Chem. Soc.* **83,** 33 (1987).
Mazur, P., and van Saarloos, W., *Physica A (Amsterdam)* **115A,** 21 (1982).
Metzner, A. B., *J. Rheol.* **29,** 739 (1985).
Mewis, J., *in* "Rheology" (G. Astarita, G. Marrucci, and L. Nicolais, eds.), pp. 149–168. Vol. 1, Plenum, New York, 1980.
Mewis, J., and Spaull, A. J. B., *Adv. Colloid Interface Sci.* **6,** 173 (1976).
Milliken, W. J., Gottlieb, M. N., Graham, A. L., Mondy, L. A., and Powell, R. L., *J. Fluid Mech.* **202,** 217 (1989).
Mondy, L. A., Graham, A. L., and Jensen, J. L., *J. Rheol.* **30,** 1031 (1986a).
Mondy, L. A., Graham, A. L., Majumdar, A., and Bryant, L. E., Jr., *Int. J. Multiphase Flow* **12,** 497 (1986b).

Mondy, L. A., Graham, A. L., Stroeve, P., and Majumdar, A., *AIChE J.* **33**, 862 (1987).
Mooney, M., *J. Colloid Sci.* **6**, 162 (1951).
Muthukumar, M., and Freed, K. F., *J. Chem. Phys.* **76**, 6195 (1982).
Nadim, A., Cox, R. G., and Brenner, H., *J. Fluid Mech.* **164**, 185 (1986).
Nguetseng, G., *J. Mec. Theor. Appl.* **1**, 951 (1982).
Nunan, K. C., and Keller, J. B., *J. Fluid Mech.* **142**, 269 (1984).
O'Brien, R. W., *J. Fluid Mech.* **91**, 17 (1979).
Overbeek, J. T. G., *Adv. Colloid Interface Sci.* **10**, 251 (1982).
Pätzold, R., *Rheol. Acta* **19**, 322 (1980).
Peterson, J. M., and Fixman, M., *J. Chem. Phys.* **39**, 2516 (1963).
Pieranski, P., *Contemp. Phys.* **24**, 25 (1983).
Pieranski, P., and Rothen, F., *J. Phys. Colloq. (Orsay, Fr)* **46** (C-3), R5 (1985).
Pusey, P. N., and Tough, R. J. A., *Adv. Colloid Interface Sci.* **16**, 143 (1982).
Pusey, P. N., and van Megen, W., *J. Phys. (Orsay, Fr.)* **44**, 285 (1983).
Rallison, J. M., *J. Fluid Mech.* **84**, 237 (1978).
Rallison, J. M., and Hinch, E. J., *J. Fluid Mech.* **167**, 131 (1986).
Reed, C. C., and Anderson, J. L., *AIChE J.* **26**, 816 (1980).
Renland, P., Felderhof, B. U., and Jones, R. B., *Physica A (Amsterdam)* **93A**, 465 (1978).
Rosensweig, R. E., *Sci. Am.* **247**, 136 (1982).
Rosensweig, R. E., "Ferrohydrodynamics." Cambridge Univ. Press, London, 1985.
Rosensweig, R. E., *Annu. Rev. Fluid Mech.* **19**, 437 (1987).
Russel, W. B., *J. Colloid Interface Sci.* **55**, 590 (1976).
Russel, W. B., *J. Fluid Mech.* **85**, 209 (1978).
Russel, W. B., *J. Rheol.* **24**, 287 (1980).
Russel, W. B., and Benzing, D. W., *J. Colloid Interface Sci.* **83**, 163 (1981).
Rutgers, I. R., *Rheol. Acta* **2**, 202 (1962a).
Rutgers, I. R., *Rheol. Acta* **2**, 305 (1962b).
Ryskin, G., and Rallison, J. M., *J. Fluid Mech.* **99**, 513 (1980).
Saito, N., *J. Phys. Soc. Jpn.* **5**, 4 (1950).
Saito, N., *J. Phys. Soc. Jpn.* **7**, 447 (1952).
Sanchez-Palencia, E., "Nonhomogeneous Media and Vibration Theory," Lecture Notes in Physics, Vol. 127. Springer-Verlag, Berlin and New York, 1980.
Sangani, A. S., and Acrivos, A., *Int. J. Multiphase Flow* **8**, 343 (1982).
Sather, N. F., and Lee, K. J., *Prog. Heat Mass Transfer* **6**, 575 (1972).
Saville, D. A., *Annu. Rev. Fluid Mech.* **9**, 321 (1977).
Schaefer, D. W., Martin, J. E., Wiltzius, P., and Canndle, D. S., *Phys. Rev. Lett.* **52**, 2371 (1984).
Schmitz, R., and Felderhof, B. U., *Physica A (Amsterdam)* **92A**, 423 (1978).
Schmitz, R., and Felderhof, B. U., *Physica A (Amsterdam)* **113A**, 90 (1982a).
Schmitz, R., and Felderhof, B. U., *Physica A (Amsterdam)* **113A**, 103 (1982b).
Schmitz, R., and Felderhof, B. U., *Physcia A (Amsterdam)* **116A**, 163 (1982c).
Schonberg, J. A., Drew, D. A., and Belfort, G., *J. Fluid Mech.* **167**, 415 (1986).
Sellers, H. S., and Brenner, H., *Physico Chem. Hydrodyn.* **11**, 455 (1989).
Slattery, J. C., *J. Fluid Mech.* **19**, 625 (1964).
Sonntag, R. C., and Russel, W. B., *J. Colloid Interface Sci.* **113**, 399 (1986).
Stauffer, D., *Phys. Rep.* **54**, 1 (1979).
Stauffer, D., "Introduction to Percolation Theory." Taylor & Francis, London, 1985.
Stimson, M., and Jeffery, G. B., *Proc. R. Soc. London, Ser. A* **111**, 110 (1926).
Stokes, G. G., *Trans. Cambridge Philos. Soc.* **9**, Part II, 8 (1851).
Takamura, K., Goldsmith, H. L., and Mason, S. G., *J. Colloid Interface Sci.* **82**, 175 (1981).
Tam, C. K. W., *J. Fluid Mech.* **38**, 537 (1969).

Thomas, D. G., *J. Colloid Interface Sci.* **20**, 267 (1965).
Tözeren, A., and Skalak, R., *J. Fluid Mech.* **82**, 289 (1977).
Vand, V., *J. Colloid Sci.* **52**, 277 (1948).
van de Ven, T. G. M., and Hunter, J. R., *J. Colloid Interface Sci.* **68**, 135 (1979).
van Diemen, A. J. G., and Stein, H. N., *J. Colloid Interface Sci.* **96**, 150 (1983).
van Megen, W., and Snook, I., *J. Colloid Interface Sci.* **53**, 172 (1975).
van Megen, W., and Snook, I., *Faraday Discuss Chem. Soc.* **65**, 92 (1978).
van Megen, W., and Snook, I., *Adv. Colloid Interface Sci.* **21**, 119 (1984).
van Megen, W., Snook, I., and Pusey, P. N., *J. Chem. Phys.* **78**, 931 (1983).
Weitz, D. A., and Oliveria, M., *Phys. Rev. Lett.* **52**, 1433 (1984).
Yoon, B. J., and Kim, S., *J. Fluid Mech.* **185**, 437 (1987).
Yoshida, N., *J. Chem. Phys.* **88**, 2735 (1988).
Ziman, J. M., "Models of Disorder." Cambridge Univ. Press, Oxford, 1979.
Zuzovsky, M., Adler, P. M., and Brenner, H., *Phys. Fluids* **26**, 1714 (1983).

OPPORTUNITIES IN THE DESIGN OF INHERENTLY SAFER CHEMICAL PLANTS

Stanley M. Englund

The Dow Chemical Company
Midland, Michigan 48667

I. Introduction

The design of chemical plants to be more inherently safe has received a great deal of attention. This is due in part to the worldwide attention to issues in the chemical industry brought on by the gas release at the Union Carbide plant in Bhopal, India, in December, 1984. A number of articles have been written on designing inherently safe plants. The purpose of this chapter is to examine and categorize some techniques that can be used by design engineers to make the plants they design inherently safer.

A major contributor to the field of literature regarding inherently safe chemical plants is Trevor Kletz, formerly with Imperial Chemical Industries, Ltd. (England), who is quoted frequently in this chapter. Kletz is now an industrial professor at the Loughborough University of Technology, Loughborough, England. Another important source of information is Frank Lees, professor of plant engineering, Department of Chemical Engineering, Loughborough University of Technology, Loughborough, England. Lees' monumental two-volume encyclopedia, *Loss Prevention in the Process Industries* (1980), has a wealth of information on quantitative methods of providing loss prevention technology in the chemical, petrochemical, and petroleum industries.

The term inherent means "belonging by nature, or the essential character of something." An inherently safe plant is safe by its nature and the way it is constructed. The term intrinsic has a meaning similar to inherent, but common usage of intrinsic in the chemical industry usually means a protection technique related to electricity. According to Lees (1980), intrinsic safety is based on the restriction of electrical energy to a level below which sparking or heating effects cannot ignite an explosive atmosphere.

There is no method of making a plant truly inherently safe, since there is always risk when human activity is involved. But, if we carefully examine the technology available to us, we can make chemical plants inherently safer than they might be without such an examination. We can determine that a plant *can* be safe, but there are many factors that will determine whether a plant *will* be safe. CEFIC, the European Council of Chemical Manufacturers' Federations (*CEFIC*, 1986), reports "these include human factors that are so difficult to quantify that they are rarely taken into consideration." They include the human side of plant management, operation, and maintenance. Designers cannot do much about these human factors, but they can often do a lot to make the plant easy to operate, and reduce the chances of accidents that may result from human error and mechanical failure.

II. Identification of Hazards

Very often, indepth safety studies come late in the design of a plant. We then try to control hazards that are identified later in the design by adding protective equipment. If hazards can be identified early in the design, changes are usually much easier and cheaper, and often better, than changes made late in the design (Kletz, 1985a). Quantitative risk assessment of the plant design can be of great value in building a safer plant as well as in improving an existing plant. There are several groups of people who can contribute expertise to the safe design of chemical plants early in the design process. These people can be divided loosely into the following categories: (1) Process designers, (2) project managers, (3) manufacturing personnel, (4) research personnel, and (5) safety and loss prevention specialists. Management is responsible for any project, but many very important decisions are routinely made by the technical experts on whom management relies. There are many decisions that can and should be made primarily by management early in the design to provide the best possible plant design. Safety and loss prevention specialists can be very helpful to the process designer at all stages in design, especially in safety reviews and quantitative risk assesssment. Research people usually contribute most of their expertise early in the design. Manufacturing personnel should be well represented at all stages of a design project.

This chapter is divided into three sections. The first section discusses general design opportunities. These primarily involve policy, layout, inventory, and process decisions having a broad impact. To make the right decision, a strong input from management is required, although other expertise, especially from safety and loss prevention experts, is required to provide information that will lead to the best possible process decision. In the second section, process design opportunities will be discussed. These require primarily the knowledge and experience of technical experts in the areas of process design, research, and manufacturing, although in some cases strong management input is required, and in all cases, strong management support is needed. For example, the design of pressure relief systems is a highly technical and advanced field, but if the latest design methods are to be of any value, management has to make a decision to devote the necessary resources to use these design methods. In the third section of this chapter, equipment design opportunities will be discussed. These opportunities require input from a variety of technical support groups as well as manufacturing experts. The experience of a successful plant operation is especially valuable here so that proven techniques can be used as much as possible.

III. General Design Opportunities

A. Clear Responsibility for Safety in Design and Operation

It is very important that the responsibility for the safe design and operation of a plant be clearly defined early in the design process. This means that competent and experienced people should be made responsible and held accountable for decisions made from the start of plant design. It is desirable to have a person or persons experienced in manufacturing operations and available in a leadership role to assist in early decisions that will affect safety. Management people should make it clear that they support and insist on process design safety. It is the responsibility of management to make certain that responsibility for safe design is clearly understood, but process design personnel should make themselves heard if they don't think this is happening.

B. Critically Review Alternatives Early in Design

Much of the rest of this chapter will discuss alternatives available to process designers. They should attempt to identify hazards early in the design, and identify alternatives available to remove or reduce them. They may then be able to remove many of the hazards or reduce their severity by making changes early in the design. The poor alternative is to add protective equipment at the end of the design or after the plant is operating, which can be expensive and not entirely satisfactory.

One constraint on the development of inherently safer plants is logistic rather than technical. Hazards and operability studies as well as other safety studies and reviews normally take place late in the design. At this stage, it is usually too late to increase design pressure of vessels, relocate electrical equipment, revise the layout, or extensively revise the process. Time must be allowed in the early stages of design for critical reviews and evaluation of alternatives. This involves an early hazards and operability (HAZOP) study using the flowsheets before final design begins (Kletz, 1985a). The use of HAZOP, fault tree analysis, checklists, audits, and other review and checking techniques can be very helpful. The techniques are extensively discussed in technical literature and will not be discussed in detail in this chapter.

C. Incorporating Emergency Planning into Original Plant Designs

Emergency planning is often done *after* the plant is nearly completed and ready for start-up. At this point, it is difficult or perhaps even possible to make some of the desired changes for emergency planning. Emergency planning should cover such items as tornado and storm shelters, flood protection, earthquakes, possible vapor releases from the proposed plant and nearby plants, proximity to public areas, accessibility of fire protection equipment to the plant, and safe exit routes. This is primarily for the protection of plant personnel and people in nearby areas who could be affected by plant problems. Emergency planning should be considered early in the design so as to make it possible to have a good plan.

D. Providing Adequate Space between Process Plants, Tanks, and Roads

When designing a safe chemical processing plant, there are few things more important to consider than the amount of space provided for the processing, storage, loading, and unloading of raw materials. There are many methods that can be used as guidelines for spacing in the design of chemical plants. Normally, it is not a good idea to have tank car and tank truck loading and unloading facilities near each other or near the process area when hazardous materials are involved. Loading and unloading facilities for hazardous materials should also not be near storage facilities. Because of the volume of chemical usually stored, the potential severity of a gas release accident from the storage of a highly toxic or flammable chemical is far greater than might occur in the process area or in the loading or unloading area. On the other hand, the frequency of storage accidents is quite low compared to loading and unloading accidents and process accidents. This poses logistics problems, which should be faced early in process layout, of how to keep storage tanks for highly toxic or flammable chemicals as far from property lines as possible, and how to maintain adequate distances between loading, unloading, storage, and process areas. In addition, it is desirable to have adequate space between the storage, loading, unloading, and process areas, and the following other areas:

(1) Other tank farms
(2) Other plants
(3) Maintenance and other facilities
(4) Power generation and other utilities
(5) Warehouses
(6) Administration and other office buildings
(7) Roads and railroads

Buffer zones should be considered to separate relatively hazardous operations from surroundings such as residential areas or other public areas. One method to obtain adequate space between loading and unloading facilities, the process area, and the storage area is shown in Fig. 1, where triangular spacing is shown. This may be suited to small sites where there are a limited number of individual plants. No facilities involving people, significant equipment, or buildings would normally be allowed inside the triangle. Such things as storage ponds and effluent treatment facilities could be in the triangle. Other facilities such as a control room, boiler, maintenance facilities, and offices would be well removed from this triangle.

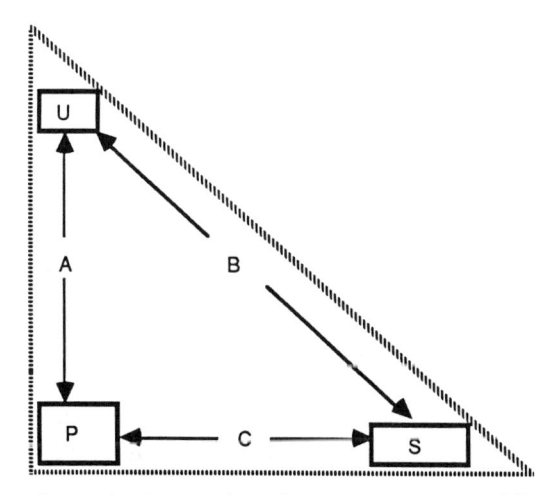

FIG. 1. Triangular spacing between the major process components. Offices, control rooms, maintenance, and similar departments should be located outside the triangle. Distances A, B, and C should be carefully selected based on the nature of the process and local conditions. U, Loading and unloading zone for hazardous materials; P, process plant; S, storage space for hazardous materials.

E. Using Minimum Storage Inventory of Hazardous Materials

The best way to minimize leaks of a hazardous or flammable material is to have less of it around. In the Flixborough disaster (Lees, 1980), on June 1, 1974, the process involved the oxidation of cyclohexane to cyclohexanone by air (with added nitrogen) in the presence of a catalyst. The cyclohexanone was converted to caprolactam which is the basic raw material for Nylon 6. The reaction product from the final reactor contained approximately 94% unreacted cyclohexane at 155°C and at over 200 psi. The holdup in the reactors was about 240,000 lbs, of which about 80,000 lbs escaped. It is estimated that $\sim 20,000-60,000$ lbs was actually involved in the explosion. The resulting large, unconfined vapor cloud explosion (or explosions, there may have been two) and fire killed 28 people and injured 36 at the plant and many more in the surrounding area. It also demolished a large chemical plant and damaged 1821 houses and 167 shops. The very large amount of flammable liquid, well above its boiling and flash point, contributed greatly to the extremely severe nature of the disaster.

In addition to the large in-process inventory of cyclohexane, there were over 430,000 gal of flammable materials, including cyclohexane and naphtha, stored at the Flixborough site. Licenses had only been issued for storage of 7000 gal of naphtha and 1500 gal of gasoline. However, the unlicensed storage of fluids had no effect on the disaster since the disaster was in the process area and not the storage area.

The results of the Flixborough investigation make it clear that the large inventory of flammable material in the process plant contributed to the scale of the disaster. It was concluded that "limitation of inventory (of flammable materials) should be taken as specific design objective in major hazard installations". Note that reduction of inventory may require more frequent and smaller shipments. There may be more chances for errors in connecting and reconnecting. These possibly negative benefits should also be analyzed. Some methods of reducing inventory of hazardous materials are discussed later in this chapter.

F. Designing Liquid Storage So Leaks and Spills Do Not Accumulate Under Tanks or Process Equipment

It is a good idea to design dikes that will not allow flammable or combustible materials to accumulate around the bottom of storage tanks or process equipment in case of a spill. If liquid spills and ignites inside a dike where there are storage tanks or process equipment, the fire may be continuously supplied with fuel and the consequences can be severe. It is usually much better

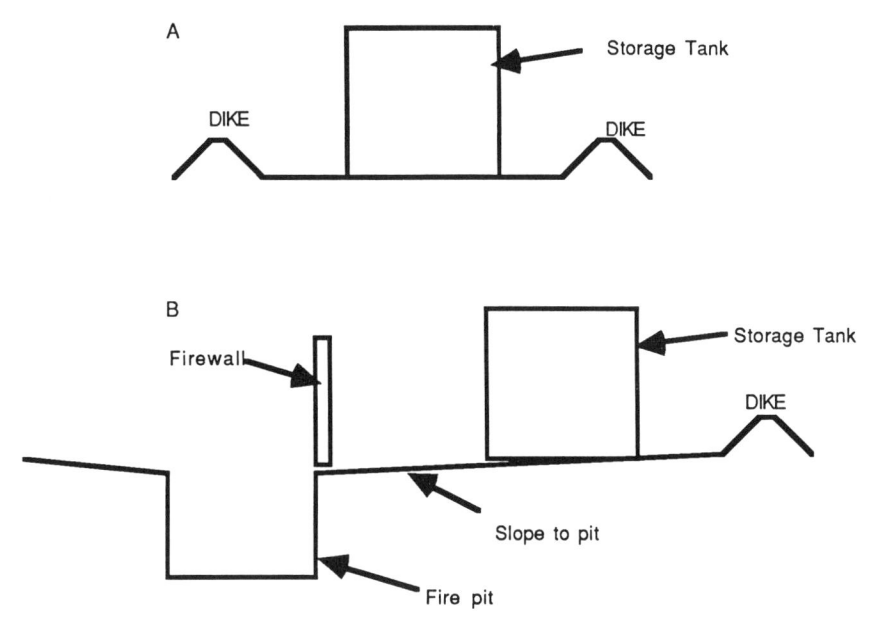

FIG. 2. Methods of diking for flammable liquids. A, Traditional diking method allows leaks to accumulate around the tank. In case of a fire, the tank will be directly exposed to flames that can be supplied by lots of fuel and will be hard to control. B, In the more desirable method, leaks are directed away from the tank. In case of fire, the tank will be shielded from most flames and fire will be easier to fight.

to direct possible spills and leaks to an area away from the tanks or equipment and provide a fire wall to shield the equipment from most of the flames if a fire occurs. Figure 2 shows schematically a traditional way to design diking as well as a better design that has met with success. Even if stored material is not flammable, allowing material to accumulate under tanks and equipment may not be desirable. For example, if bromine is spilled in a nondrained dike area containing bromine storage tanks, the automatic dump valve on the tanks may rapidly become corroded on the outside making it impossible to transfer contents to another storage vessel.

The design in Fig. 2A is usually undesirable for flammable liquids because it will allow flames from an ignited spill within the diked area to "cook" the tank. Such a fire may be very dangerous and hard to control because of the possibility of rupturing the tank. The better design in Fig. 2B will divert spills from the immediate area of the tank. In the event of a fire, the tank will be shielded by the fire wall. This safeguards the tank from direct exposure to the flames. Such a fire is easier to control, and the tank is less likely to rupture.

G. USE OPEN STRUCTURES FOR PLANTS USING FLAMMABLE OR COMBUSTIBLE MATERIALS

There are many examples of serious fires and explosions that probably resulted in part from handling moderate to large quantities of flammable or combustible liquids and liquefied flammable gases inside enclosed structures. If a sufficient quantity of flammable mixture should ignite inside an ordinary chemical processing building, it is highly probable an explosion will occur that will seriously damage the building. For this reason, processing equipment is often installed in a structure without walls, usually called an "open structure." This permits effective ventilation by normal wind currents and aids the dispersion of any vapors that escape. If gas ignites in the structure, the absence of walls minimizes the pressure developed from the combustion and the probability of flying shrapnel from a shattered structure (Howard and Karabinis, 1981). Substantial damage can be done to a building by the combustion of a surprisingly small quantity of a flammable gas–air mixture. If there is an explosion in a building where the flammable gas mixture occupies a space equal to only 1 or 2% of the building volume, the building may be seriously damaged if it does not have adequate explosion venting. This results because most buildings can suffer substantial structural damage from an internal pressure appreciably less than 1 psi (0.07 bar). Thus, a building does not need to be full or even close to full of a flammable mixture for a building explosion to occur that can cause considerable damage. This is not just theoretical, it has been proven in real life experience (Howard and Karabinis, 1981)!

Brasie (1976) reported on several hazard levels of interest for fires in enclosed or semiconfined spaces. For a fuel in vapor form that has a heat of combustion of 19,000 BTU/lb, the following weights of fuel enclosed in a 1000 ft^3 building are required to achieve the effects shown, assuming no venting.

Weight of Fuel (lbs).	Effect
0.15–0.2	Pressure reaches 1 psig: Significant building damage
0.18	Temperature reaches 125°C: Maximum personnel tolerance
0.25	Significant hazard
0.6	Temperature reaches 425°C: Cellulose ignites
3–	Room engulfed in flames

In 1950, a serious explosion occurred in Midland, MI at The Dow Chemical Company in a chemical processing unit. Butadiene and styrene, in vapor and liquid forms, were accidentally discharged into a large processing area. The building was 130 × 288 ft and one story high with some steel decks. All electrical equipment in the area was explosion proof, and there was adequate automatic fire sprinkle protection. The walls were brick, and the roof was precast concrete supported on exposed steel trusses. The leak occurred in a 50 × 130 ft section of the building which was separated from the rest of the building by a 12-in bonded brick firewall with one protected opening. The explosion caused the following events to occur:

(1) Of the 40 men in the building, eight were killed.
(2) Eighty percent of the concrete slabs were blown off the roof.
(3) Walls around the 50 × 130 ft section, including the firewall, were completely demolished.
(4) All piping and duct work was completely demolished.
(5) Process vessels were reduced to salvage value.
(6) Relatively minor damage was done to buildings within a radius of 1000 ft by flying debris and concussion waves.
(7) The entire department was out of operation for approximately six months.

This tragedy was instrumental in causing Dow to establish a policy of using open structures for chemical processes that use substantial quantities of flammable liquids and liquefied flammable gases.

H. Avoid Buried Tanks

At one time, burying tanks was recommended because this minimized the need for fire protection systems, dikes, and distance separation. At Dow, this is no longer considered good practice except in special cases. Mounding or burying tanks above grade has most of the same problem as burying tanks below ground and is usually not recommended. There are several problems with buried tanks.

(1) Difficulty in monitoring exterior corrosion and shell thickness
(2) Difficulty in detecting leaks
(3) Difficulty repairing tanks if the surrounding earth is saturated with chemicals from a leak
(4) Potential groundwater contamination from leaks

Governmental regulations concerning buried tanks are becoming more and more strict. This is because of the large number of leaking tanks that have been identified as causing adverse environmental and human health problems.

The Environmental Protection Agency's (EPA) office of underground storage tanks defines underground tanks as those with 10% or more of their volume underground, including piping. According to an article by Brooks and others (1986), the EPA found that about 65% of documented leak incidents involved retail gasoline stations and only 3% involved chemicals (as defined by the EPA). About 4% of underground tanks are believed to contain commercial chemical products. Leaks were caused by the following failures, which are based on a survey of 12,444 tanks: (1) Structural failure, 46%; (2) corrosion, 26%; (3) loose fittings, 14%; (4) improper installation, 7%; and (5) natural phenomena, 6%.

Eleven states have enacted laws setting standards for underground storage tanks. The EPA (1987) has issued regulations requiring notification to the appropriate regulatory agencies about age, condition, and size of tanks that store hazardous wastes and underground storage tanks containing commercial chemical products. Final technical requirements have been promulgated for hazardous waste storage tanks. At the time of this writing, technical requirements for underground storage tanks containing regulated substances have been proposed by the EPA. Florida reportedly requires owners, as an early detection measure, to install at least four monitoring wells around a buried tank area. It is estimated that it will cost $150,000–$250,000 per site to clean up leaking buried tanks in Florida.

Small leaks are difficult to detect. The National Fire Protection Association of Quincy, MA, has established 0.05 gal/hr as the rate above which a tank is considered to be leaking. Leak detection measurements can be influenced by many factors, making it difficult to detect small leaks. Some of these factors are changes in product temperature and pressure, vapor pockets, tank geometry and inclination, surface waves, water table, and operator error. The EPA (1987) has proposed new rules to prevent leaks from underground tanks. The rules require testing for leaks within three years and gradually end the use of leak-prone tanks within 10 years. Leaks detected in testing would have to be repaired immediately. The owners or operators would be responsible for any damage to people or the environment and for replacing tanks. All new petroleum tanks would have to be protected from corrosion and have a leak-detection apparatus and devices to contain overflow and spills. Tanks storing chemicals would be required to have double walls or be placed in a concrete vault or lined excavation.

Because of more stringent regulatory requirements and potential future liabilities associated with buried tanks, it is probably inherently safer to use above-ground open storage with suitable spacing, diking, and fire protection facilities. This will avoid most of the potential groundwater contamination problems that could result from a leak. The potential problems from leaks can be exceedingly expensive and harmful to the environment if buried tanks are improperly designed, constructed, and maintained. Some states, and parts of some countries, require flammable liquid storage tanks to be buried unless a modification is granted. Local authorities should be contacted to check legal requirements.

I. Constructing Process and Storage Areas Away from Residential or Potentially Residential Areas

The Bhopal, India plant of the Union Carbide Corp. was originally built 1.5 miles from the nearest housing; but with time, a shanty town grew up next to the plant. This demonstrates the need to prevent hazardous plants from being located near residential areas (Kletz, 1985a). If possible, the cost of a plant should include funds for an adequate buffer zone unless other means can be provided to ensure that the public will not build adjacent to the plant. The nature and size of this buffer zone depends on many factors, including the amount and type of chemicals stored and used. If the land required to separate the plant from public areas can be put to some other use, such as plants with low hazards, laboratories, or light industry, and maintain a low population density, a satisfactory buffer may be provided. This assumes that personnel in the buffer area can be protected or evacuated if necessary. If possible, the land should be under the control of the plant, or there should be assurance from credible sources that people in the buffer zone will not be allowed near the plant. Where feasible, zoning restrictions may provide an alternative to land purchase. A possible negative effect of increased distances from populated areas could be an increase in emergency-help response time. This should be considered when the overall effect of plant location is being reviewed.

J. Designing for Total Containment

In general, it is safer to totally contain, or nearly totally contain, hazardous and flammable materials in chemical processes, if it is reasonably practical to do so, than to allow these materials to escape into the environment. In many cases, this can be accomplished by designing the processing equipment to withstand the maximum pressure that can be expected from runaway polymerizations or other reactions or explosions. This requires detailed knowledge of the process and the possible overpressure that could result. This knowledge can best be obtained from experimental data combined with a theoretical analysis.

It is helpful if management establishes the goal of total containment early in the process design so that all those involved can contribute their part to achieving that goal. For example,

(1) Explosions involving organic dusts can usually be contained in equipment that can withstand about 7–10 times the initial pressure (Lees, 1980). Thus, if the system is originally at atmospheric pressure, the maximum pressure to be expected is about 100–150 psig. Data on explosion characteristics of the actual material to be processed are usually necessary to confirm the required design pressure.

(2) Dikes around the process and storage areas with only controlled pump discharge from the diked areas will make controlled liquid release from spills possible. The diked areas must not be allowed to fill with water, snow, or ice.

(3) Massive vapor releases can be minimized by choosing the proper design pressure for processing equipment and storage vessels.

(4) The design of emergency relief systems is important. The goal should be well-designed systems that will relieve pressure safely but not discharge material uncontrollably (as single frangible disks often do). The use of flares, incinerators, scrubbers, and emergency recovery tanks should be considered.

(5) Construction materials should be selected that will withstand long-term exposure to the chemicals used in the process. This includes gaskets, which tend to be weak points in piping and vessels. Gaskets are discussed in Section V,E of this chapter. Occasionally fire-safe emergency valves are installed with gaskets that will quickly burn away if there is a fire. This tends to defeat the purpose of fire-safe valves.

K. Redesigning Obsolete Plants Before Accidents Occur

Occasionally because of job rotation, promotion, retirement of key individuals, poor business, or just plain poor management, a "negative learning curve" occurs in a process plant. Process designers should do all they can to ensure that all existing process plants are operated safely. They may see things that are not apparent to those actually running the plant. Process designers should not be reluctant to point out the shortcomings of plants and suggest realistic corrective measures commensurate with the economic realities of the situation. At Dow there is a saying that you can "expect what you inspect." Management should support the idea of regular audits of plants so that existing hazards can be identified and either controlled or eliminated.

The Flixborough disaster in June, 1974 (Lees, 1980), is an example of a case where a modification was introduced into a mostly well designed and constructed plant. This modification destroyed the plant's integrity and contributed to a major accident. The modification was made when a reactor failed (a large crack had formed). The modification was inadequate and the remaining reactors were not examined.

L. Storing Liquefied Gases at Low Temperature and Pressures

Usually, leaks of liquefied gases are less serious if the liquefied gases are refrigerated at low temperatures and pressures than if they are stored at ambient temperatures under pressure. A leak of a volatile liquid held at atmospheric temperature and pressure results in only a relatively slow evaporation of the liquid. Escape of a refrigerated liquefied gas at atmospheric pressure gives some initial flashoff followed by an evaporation rate that is relatively slow but faster than the first case, depending on weather. Loss of containment of a liquefied gas under pressure and at atmospheric temperature, however, causes immediate flashing of a large proportion of the gas. This is followed by slower evaporation of the liquid residue and is usually the most serious case. The hazard from a gas under pressure is normally much less in terms of the amount of material stored if it is not liquefied, but the physical energy released is large if a confined explosion occurs at high pressure.

The economics of storing liquefied gases are such that it is usually attractive to use pressure storage for small quantities, pressure or semirefrigerated storage for medium to large quantities, and fully refrigerated quantities for very generally considered that there is a greater hazard in storing large quantities of liquefied gas under pressure than at low temperatures and low pressures. The trend is toward replacing pressure storage with refrigerated low pressure storage for large inventories. However, this is not always the case. It is necessary to consider the risk of the entire system, including the refrigeration system, and not just the storage vessel. The consequences of a failed refrigeration system must be considered. Lees (1980) states that the Imperial Chemical Industries Liquefied Flammable Gas (ICI LFG) code recommends the separation distance between storage and an ignition source to be, for ethylene, 60m for pressure storage and 90m for refrigerated low pressure storage, and for C_3s, 45m for both types of storage. The general approach taken by ICI is that there is a significant risk of failure for refrigerated storage tanks but a negligible risk for pressure storage vessels. Each case should be carefully evaluated on its own merits. In most cases, refrigerated storage of hazardous liquefied gases is undoubtedly safer, such as in the storage of large quantities of liquefied chlorine (Lees, 1980).

IV. Process Design Opportunities

The knowledge and experience of technical experts are important in identifying process design opportunities of inherently safer chemical plants. Management may or may not be familiar with the specific design opportunities that exist to make plants inherently safer. The process designer should make sure these design opportunities are adequately considered and put into practice whenever possible. Occasionally it is necessary for the process designer to point out problems and the possible benefits that could result from their solution. Occasionally it will be necessary to ask for additional resources, such as additional technical help to design pressure relief systems properly, in order to take advantage of the opportunities that will appear. A discussion of specific process design opportunities follows.

A. Understanding the Reactive Chemicals and Reactive Chemical Systems Involved

The main business of most chemical companies is to manufacture products through the control of reactive chemicals. The reactivity of chemicals that makes them useful can also make them hazardous. Therefore, it is essential that process designers understand the nature of the reactive chemicals involved in the process (*Corporate Safety*, 1981). Usually reactions are carried out without mishaps, but occasionally chemical reactions get out of control because the wrong raw materials were used, operating conditions were changed, unanticipated time delays occurred, or equipment failed. A chemical plant can be inherently safer if we use knowledge of the reactive chemicals systems in the plant's design. We must understand the chemistry of the process if we are to design a safe chemical processing plant.

1. *Reactive Hazard Evaluations*

Reactive hazard evaluations should be made on all new processes as well as existing processes on a periodic basis. There is no substitute for experience, good judgement, and good data when evaluating potential hazards. Reviews in process chemistry should include (a) reactions, (b) side reactions, (c) heat of reaction, (d) potential pressure buildup, and (e) intermediate streams. Reactive chemicals test data should be reviewed for evidence of flammability characteristics, exotherms, shock sensitivity, or other evidence of instability. Examine Planned operation of the process should be examined, especially for

(a) upsets,
(b) modes of failure,
(c) delays,
(d) redundancy,
(e) critical instruments and controls, and
(f) worst credible case scenarios.

2. *Worst Case Thinking*

At every point in the operation, the process designer should conceive of the worst possible combination of circumstances that could *realistically* exist such as

- Loss of cooling water
- Power failure
- Wrong combination or amount of reactants
- Wrong valve position
- Plugged lines
- Instrument failure
- Loss of compressed air
- Air leakage
- Loss of agitation
- Deadheaded pumps
- Raw material impurities

An engineering evaluation should then be made of the worst case consequences with the goal that the plant will be safe even if the worst case occurs. When the process designers know what the worst case conditions are, they should try to avoid worst case conditions, be sure adequate redundancy of safety systems exists, and identify and implement lines of defense. These lines of defense could be preventive measures, corrective measures or sometimes as a last resort, containment or possibly abandonment of the process if the hazard is unacceptable. It is important to note that the worst case should be something that is realistic, not something that is conceivable but extremely unlikely.

Dow has adopted the following philosophy for design scenarios in terms of independent causative effects:

(1) All single events that can actually and reasonably occur are credible scenarios.

(2) Scenarios that require the coincident occurrence of two or more totally independent events are not credible design scenarios.

(3) Scenarios that require the occurrence of more than two events in sequence are not credible.

(4) A failure that occurs while an independent device is awaiting repair represents but one failure during the time frame of the initiation of the emergency and is therefore credible. The lack of availability of the unrepaired device is a pre-existing condition.

3. *Reactive Chemicals Testing*

Much reactive chemical information involves thermal stability and the determination of the temperature at which an exothermic reaction starts, the rate of reaction as a function of temperature, and the heat generated per unit of material. Following is a review of some of the main sources of reactive chemicals data.

a. *Calculations.* Potential energy that can be released by a chemical system can often be predicted by computerized thermodynamic calculations. If there is little energy, the reaction still may be hazardous if gaseous products are produced. Kinetic data is usually not available in this way. Thermodynamic calculations should be backed up by actual tests.

b. *Differential Scanning Calorimetry (DSC).* Sample and inert reference materials are heated in such a way that the temperatures are always equal. If an exothermic reaction occurs in the sample, the sample heater requires less energy than the reference heater to maintain equal temperatures. If an endothermic reaction occurs, the sample heater requires more energy input than the reference heater. Onset-of-reaction temperatures reported by the DSC are higher than the true onset temperatures, so the test is mainly a screening test.

c. *Differential Thermal Analysis (DTA).* Sample and inert reference materials are heated at a controlled rate in a single heating block. If an exothermic reaction occurs, the sample temperature will rise faster than the reference temperature. If the sample undergoes an endothermic reaction or a phase change, its temperature will lag behind the reference temperature. This test is basically qualitative and can be used for identifying exothermic reactions. Like the DSC, it is also a screening test. Reported temperatures are not reliable enough to be able to make quantitative conclusions. If an exothermic reaction is observed, it is advisable to conduct tests in the Accelerating Rate Calorimeter (ARC[1]).

d. *Accelerating Rate Calorimeter.* This method determines the self-heating rate of a chemical under near-adiabatic conditions. It usually gives a conservative estimate of the conditions for, and consequences of, a runaway reaction. Pressure and rate data from the ARC may sometimes be used for pressure-vessel emergency-relief design. Activation energy, heat of reaction, and approximate reaction order can usually be determined. For multiphase reactions, agitation can be provided. The ARC can provide extremely useful and valuable data. An example of data from an ARC run is shown in Fig. 3. The Vent Sizing Package (VSP), which was recently developed, is an extension of ARC technology. The VSP is a bench scale apparatus for characterizing

---

[1] ARC is a trademark of Columbia Scientific Industries Corp.

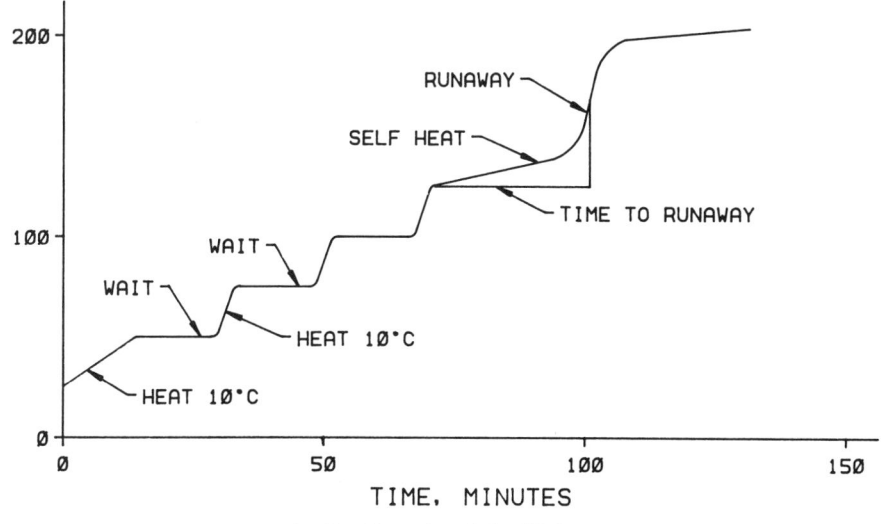

FIG. 3. Operation of the ARC.

runaway chemical reactions. It makes possible the sizing of pressure relief systems with less engineering expertise than is required with the ARC or other methods. The VSP is discussed in Section V,C of this chapter.

e. *Shock Sensitivity.* Shock sensitive materials react exothermally when subjected to a pressure pulse. Materials that do not show an exotherm on a DSC or DTA are presumed not to be shock sensitive. Three testing methods are listed here.

(1) Drop Weight Test—A weight is dropped on a sample in a metal cup. The test measures the susceptibility of a chemical to decompose explosively when subjected to impact. Weight and height can be varied to give semiquantitative results for impact energy. This test should be applied to any materials known or suspected to contain unstable atomic groupings.

(2) Confinement Cap Test—Detonability of a material is determined using a blasting cap.

(3) Adiabatic Compression Test—High pressure is applied rapidly to a liquid in a U-shaped metal tube. Bubbles of hot compressed gas are driven into the liquid and may cause explosive decomposition of the liquid. This test is intended to simulate "water hammer" and "sloshing" effects in transportation such as humping of railway tank cars. It is very severe and gives worst case results.

f. *Flammability: Flash Point.* The "closed cup" flash point determination produces the most important data for determining the potential for fire. The flash point is the lowest temperature at which the vapors can be ignited under conditions defined by the test apparatus and method.

g. *Flammability: Flammable Limits.* Flammable limits, or the flammable range, are the upper and lower concentrations (in volume percent) which can just be ignited by an ignition source. Above the upper limit and below the lower limit, no ignition will occur. Data are normally reported at atmospheric pressure and at a specified temperature. Flammable limits may be reported for atmospheres other than air, and at pressures other than atmospheric.

h. *Flammability: Autoignition Temperatures.* The autoignition temperature of a substance, whether liquid, solid, or gaseous, is the minimum temperature required to initiate self-sustained combustion in air with no other source of ignition. Ignition temperatures should be considered only as approximate. Test results tend to give temperatures higher than the actual autoignition temperature.

i. *Dust Explosions.* Combustible, dusty materials, with particle sizes less than ~200 mesh, can explore if a sufficient concentration in air is present along with an ignition source. The standard test has been designed to determine rates of pressure increase during an explosion, the maximum pressure reached, and the minimum energy needed to ignite the material. These data are useful in the design of safe equipment for handling dusty combustible materials in a process. One test apparatus widely applied is the Hartmann Tube which has a volume of 1.3 1. However, Bartknecht (1981) emphatically states "test results measured in the Hartmann Apparatus underrate the effects of dust explosions and are not a suitable basis for design of protective measures." Combustible dusts need a minimum volume to develop their full reaction velocity. Bartknecht states that to determine explosion data for combustible dusts, a minimum volume of 16 1 is required to ensure correlation with data from large test vessels. This has been confirmed by comprehensive testing with a 20 1 sphere (Kletz, 1985b).

B. REDUCING INVENTORY BY CHANGING PROCESS CHEMISTRY

If given enough time to develop new processes, research chemists and chemical engineers can often devise new processing techniques that will reduce the hazard of a process by reducing the maximum instantaneous inventory of hazardous materials. Occasionally these new techniques also provide increased productivity, quality, and improved properties. For example, in the past, synthetic rubber latex made from styrene and butadiene was usually made by adding all the ingredients to a batch reactor or a series of continuous reactors and carrying out the reaction. For small and specialty uses, it is not always practical or desirable to make these latexes by a continuous process. Indeed, some of the continuous processes for latex manufacturing have so much unreacted monomer present, they may be less safe than a conventional batch process. In conventional batch processes for making latex, all the reactants are added at the beginning of the reaction. There is a considerable amount of hazardous, flammable material present in the reactor at the beginning of the reaction, and many scenarios such as leaks, agitation failure, loss of cooling, and runaway polymerization could cause safety problems.

Methods have been developed for improving batch process productivity in the manufacturing of latex. One is the continuous addition of reactants so that the reaction takes place as the reactor is being filled (Englund, 1982). These are not continuous processes even though the reactants are added continuously during most of a batch cycle. The net result is that reactants can be added about as fast as heat can be removed. There is relatively little hazardous material in the reactor at any time because the reactants, which are flammable or combustible, are converted to nonhazardous and nonvolatile polymer almost as fast as they are added.

An example is given by U.S. Patent 3,563,946 (1971). In this example, styrene–butadiene latex is prepared by simultaneously feeding two separate streams (one a mixture of styrene and butadiene monomers, the other an aqueous solution of sodium lauroyl sulfate as emulsifier, and sodium persulfate as initiator) to a reactor that has an initial charge of water and emulsifier. The reactants are added at about 25°C and the reaction occurs at 90°C. A heat balance on the reactor jacket is usually used to make certain the monomers are reacting about as fast as they are added. Also, the charging rate of reactants is limited by suitable choices of pumps, orifices, and piping, which further reduce the possibility of monomers being added too rapidly. This is done to make sure the concentration of unreacted monomers in the reactor is not building up. This could possibly cause an unexpected "bomb" in the reactor, which would then resemble a conventional batch process, but with poor control.

Compared to a conventional batch reactor, there is less hazardous material in the reactor at any time, and the consequences of power loss, loss of temperature control, leaks, and runaway reactions are much less severe. In addition, it is easier to have good temperature control, which can provide a more reproducible, high quality product. Productivity is significantly better than in conventional batch processes since the reactor is polymerizing at its maximum rate most of the time and is normally limited only by heat transfer

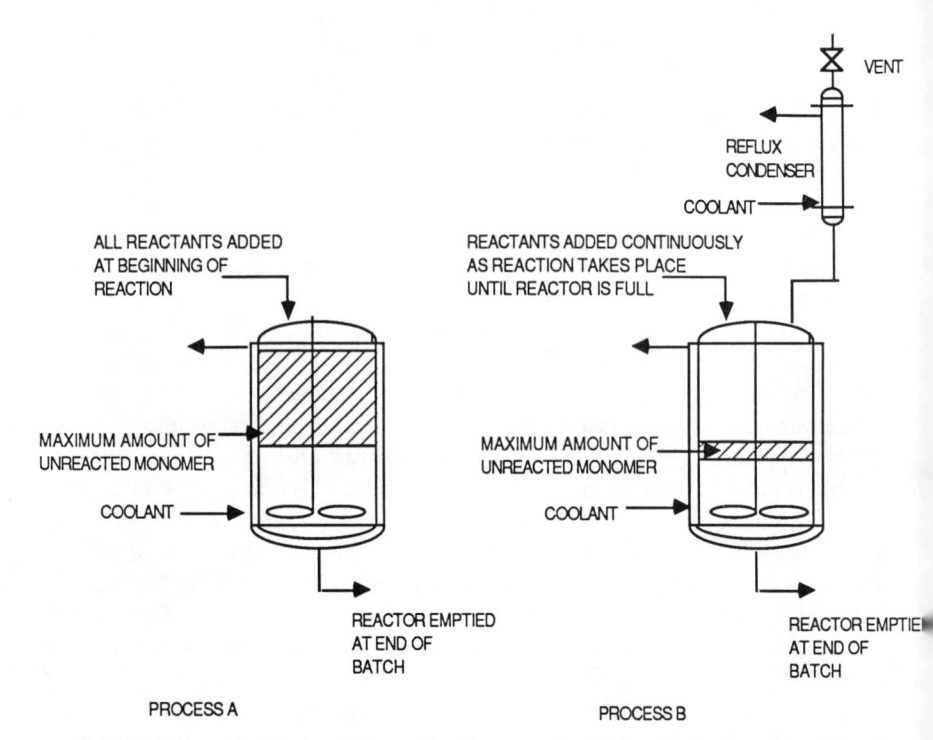

FIG. 4. Process A, Batch reaction with all reactants added at beginning of reaction. A considerable amount of flammable and hazardous material is in the reactor at the beginning of the process. Process B, a batch reaction with the reactants added during the reaction. Little flammable and hazardous material is present at any one time. Reflux (or Knockback) condensor is used to provide additional heat transfer.

capability. Heat transfer capability can be increased by adding a reflux or knockback condenser (U.S. Patent, 1977), which, in effect, provides additional heat transfer for the reaction, and consequently makes it possible to handle fast reactions or fast feed rates safely as long as there is sufficient volatile material to vaporize and be condensed. This process is shown in Fig. 4B, where it is compared to a more conventional batch process in Fig. 4A in which all the reactants are added at the beginning. Care must be taken so that when processes with less inventory are introduced, inadequately tested processes with unrecognized health, safety, and environmental risks do not result.

C. Reducing Inventory by Changing Mixing Intensity

Lees (1980) describes reduction in process inventory in the evolution of the processes for manufacturing nitroglycerine. The first is a batch process with a holdup of about 1000 kg. The second is a continuous process with a holdup of 200–300 kg. The third is a continuous process with reaction taking place in the nozzle and with a holdup of only about 5 kg. Occasionally an opportunity will arise in which a fast reaction can be made to take place in an inline pipe mixer, which has small inventory, instead of a tank with much larger inventory. The reactor volume required for continuous polymerization can sometimes be reduced by replacing a cascade of stirred tanks in series with a plug-flow tubular reactor (Levenspiel, 1979). Static mixers can sometimes be used as polymerization reactors, instead of stirred reactors or plug-flow tubular reactors, because they induce plug flow and create good radial mixing at low shear rates with low energy consumption.

Commercially available static mixer reactors (SMR) as large as 1.8 m in diameter are operating in the production of nylon, silicones, polystyrene, polypropylene, and other polymers. Removing heat from a highly exothermic reaction in a static mixer equipped with a simple cooling jacket is limited by heat transfer to small diameters, typically less than 6 in. A new type of static mixer has been developed to overcome laminar heat-flow transfer problems. It is called the SMR mixer–heater exchanger–reactor (Mutsakis et al., 1986).

The SMR has mixing properties and plug flow similar to other static mixers. However, it has hollow crossing tubes that form the shape of the mixing elements. These hollow tubes carry heat-transfer fluid, greatly increasing heat-transfer surface area and making a virtually isothermal reaction possible. Such a system has the potential to be smaller than stirred tanks in series and has no internal moving parts, such as seals, that could fail.

D. Using Low Inventory in Distillation Processes

Distillation columns have a large inventory in the reboiler, and typically, an inventory several times greater in the column itself. Column holdup may be reduced by using low holdup internals. Conventional trays and packings differ by a factor of about 10 in inventory per theoretical plate (Kletz, 1985a). An estimate of holdup per theoretical plate is shown (Lees, 1980).

Intervals	Holdup/theoretical plate
Conventional trays such as sieve, valve, etc.	40–100 mm liquid
Packing	30–70 mm liquid
Film trays	10–20 mm liquid

High efficiency packing, such as oriented wire packing (Koch/Sulzer, Flexipack, Goodlow), has a thin film of liquid on the wires so that the total liquid holdup is relatively low. Other suggestions from Lees (1980) include using tall thin columns such as those used for heat sensitive materials, and

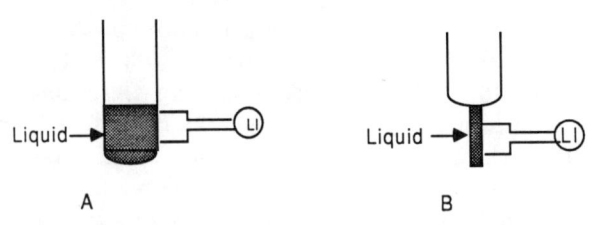

FIG. 5. A, Level indicator on the bottom of a distillation column; considerable liquid inventory. B, Level indicator on pipe leading from bottom of distillation column; low liquid inventory.

locating peripheral equipment, such as reboilers and bottoms pumps, inside the column.

A new distillation process using centrifugal force, called the Higee distillation process, is said to be able to provide a 1000-fold reduction in inventory (Kletz, 1985a). One simple method of reducing volume of liquid in the bottom of a distillation column is shown in Fig. 5B (Geyer, 1986). The inventory is reduced because the level control is used on a pipe leg on the bottom of the column, instead of using the entire column diameter as shown in Fig. 5A. This poses some control problems, but it has been shown, in practice, that they can be solved. It is often possible to reduce the inventory in distillation processes, although occasionally at somewhat increased costs. In the design of inherently safer plants, the options available to reduce inventory in distillation equipment should be carefully considered when flammable and toxic materials are involved, even though the apparent first costs may be greater.

E. MINIMIZING INVENTORY IN HEAT EXCHANGERS

Heat exchangers fail in pressure service more often than pressure vessels, largely because of mechanical and corrosion problems (Lees, 1980). Failures that occur often are the result of exposure to operating conditions more severe than those for which the system was designed. The process designer should try to anticipate unusual conditions to which heat exchangers may be subjected and install safeguards or other features to prevent failure. Failures that can occur include

(1) excessive stress by external loads or uneven tightening of flanges,
(2) stress cycling caused by pressure changes,
(3) thermal shock and thermal cycling,
(4) hydrogen attack, and
(5) corrosion failure.

TABLE I
Surface Compactness of Heat Exchangers

Type of exchanger	Surface compactness, M^2/M^3
Shell and tube	70–150
Plate	120–225
Spiral plate	up to 185
Shell and finned tube	up to 3300
Plate–fin	up to 5900
Regenerate—rotary	up to 6600
Regenerative—fixed	up to 15,000 (may be much less)

Since failure of heat exchangers can be a significant problem, it is desirable to minimize the inventory of hazardous materials in heat exchangers. Kletz (1985a) has reported on the classification of heat exchangers according to the ratio of heat-transfer surface to volume. Some typical figures are reported in Table I. Plate exchangers require a large amount of gasketing, which makes them unsuitable in some applications. Some of the difficulties seem to have been overcome since plate exchangers are now used extensively on off-shore oil platforms and increasingly in the oil industry. Plate heat-exchangers may fit into any flowsheet where heat may be exchanged between fluids at pressures and temperatures of up to 300 psi and 150°C, respectively. (Sjoren and Grueiro, 1983). Heat transfer in shell and tube exchangers can be improved, and thus the inventory reduced, by inserting a matrix of wire into the tubes, which promotes turbulence near the walls (Kletz, 1985b).

It may be possible to reduce the inventory of hazardous materials in shell and tube heat-exchangers by using careful design that results in a heat exchanger no larger than is necessary for the job. This also makes good sense from an economic point of view. Several heat-exchanger computer design programs are available that make close design possible easily and quickly. One of these programs that has received good acceptance is B-JAC[2] (Pase, 1986). This is an integrated, interactive program that can analyze in minutes dozens of alternatives that would require weeks to do by skilled individuals with hand calculators.

It also may be possible to reduce heat-exchanger inventory by using smaller tubes than usual. Manufacturing people generally don't like small tubes in heat exchangers because they are harder to clean than large tubes. Also, higher

[2] B-JAC is a registered trademark of B-JAC Computer Services, Inc., Midlothian, VA.

pressure drop may be required, which would cause higher power costs. Tubes with a diameter of $\frac{3}{4}$ in., instead of the more common 1-in. tubes, may be a reasonable compromise in many cases.

F. Reducing the Possibility of Losses from Dust Explosions (Brasie, 1986)

Most organic solids, metals, and some combustible inorganic salts can form explosive dust clouds. In order to have a dust explosion, certain elements are necessary: (1) Particles of dust of suitable size, (2) A sufficient source of ignition energy, (3) A concentration of dust within explosive limits. If an explosive dust in air that meets the above criteria occurs in a process, an explosion should be considered inevitable. The process designer of inherently safer plants must take into account the possibility of dust explosions and design accordingly.

In dust explosions the combustion process is very rapid. The flame speed is high compared to that in gas deflagrations. However, detonations normally do not occur in dust explosions in industrial plants. In a serious industrial dust explosion, two steps often occur. First, a primary explosion occurs in part of a plant, causing an air disturbance. Second, the air disturbance disperses dust and causes a secondary explosion. The secondary explosion is often more destructive than the primary explosion. There is a great deal of literature on dust explosions, which is available to the process designer. See, for example, Lees (1980) for a bibliography on dust explosions.

If, in a process, flammable (explosive) dust is inevitable, several alternatives or combinations of alternatives are available:

(1) Containment (maximum pressure is usually below 120 psig)
(2) Explosion venting to a safe place
(3) Inerting (most organic dusts are nonflammable in atmospheres containing less than about 10% oxygen)
(4) Suppression (usually a last resort)

A fundamental solution to the dust explosion problem is to use a wet process so that dust suspensions do not occur at all. If a wet process can be used, it is one of the most satisfactory methods. However, the process must be wet enough to be effective. Some dusts with a high moisture content can still ignite. Dust concentrations in major equipment can be designed below the lower flammable limit, but this often cannot be counted on in operation. Dust concentrations cannot be safely designed to be above an upper flammable limit, because such a limit is ill-defined (Lees, 1980). For a large number of flammable dusts, the lower explosion limit lies between 20 and 60 g/m^3. The upper explosion limit is in the range of 2000 and 6000 g/m³, but this number is of limited importance. A small amount of flammable gas or vapor mixed in with a flammable dust can cause an explosive mixture to be formed even if both are at concentrations below the explosive range by themselves.

Containment is possible by using pressure vessels that have a design pressure at least seven times the normal operating pressure. Since the test pressure is usually 1.5 times the design pressure, testing ensures a vessel that will not fail under the normal maximum explosion pressure of $\sim 7-10$ times the starting pressure (Bartknecht, 1981). For example, for a system operating at or near atmospheric pressure, a design pressure of ~ 100 psig will contain most dust explosions without rupturing. Care must be taken that an explosion in the contained area does not pass into other parts of the equipment that are not as well protected. Properly designed quick-closing valves can stop dust explosions from passing into downstream or upstream equipment.

Venting is only suitable if there is a safe discharge for the material vented. Whenever an explosion-relief venting device is activated, it may be expected that a tongue of flame containing some unburned dust will first be ejected. The unburned dust will be ignited as it flows out the vent and can produce a large fireball that will extend outward, upward, and downward from the vent. It is essential for protection of personnel that venting is to an open place not used by people. If a duct must be used, the explosion pressure in the enclosure will be increased considerably. Therefore, particular attention must be paid to the design of the enclosure in which the explosion could take place. Dust explosion venting was formerly expressed traditionally as a vent area per unit volume of space protected. The National Fire Protection Association 68 (NFPA) (1988) guide has nomographs that can be used to select relief areas required for combustible dusts when test data on the dusts are available. The Nomographs in NFPA 68 are by far the preferred way to design dust explosion-relief devices. A quick but less accurate method, which can be used as a rough guide, is to use the following approximation for vent area (Lees, 1980) for volumes measuring up to 1000 ft³:

Maximum rate of pressure rise ($lb_f/in.^2$ sec)	Vent ratio ft^2/ft^3
$\langle < 5000 \rangle$	1/20
$\langle 5000-10,000 \rangle$	1/15
$\langle > 10,000 \rangle$	1/10

Relief venting to reduce dust explosion pressure requires that the equipment to be protected has a certain minimum strength. If the enclosure strength is too low, the enclosure will be damaged or destroyed before the explosion relief device can function. NFPA 68 (1988) states that the strength of the enclosure should exceed the vent relief pressure by at least 0.35 psi. For industrial equipment such as dryers and baghouses, it is often desirable to have considerably more strength built into the structure to reduce the size of the vent area required. Also, the supporting structure for the enclosure must be strong enough to withstand any reaction forces developed as a result of operating the vent. An example of industrial equipment designed for explosion venting is a baghouse sold by the Dust Control Equipment Co., Jeffersontown, KY. This baghouse, designed for explosive dust being handled in a 13,500 cfm air stream, normally operates at 6 in. negative water pressure. The unit is designed for 13,500 scfm, 1 psig negative water pressure, 2 psig positive design pressure, and an instantaneous pulse pressure of 10 psig. It is constructed of four modules, each having an explosion vent membrane that ruptures at 1.5 psig. The explosion vent membranes are on the "dirty" sides of the bags, and are located on the lower, vertical exterior walls to reduce internal pressure drop in the event of an explosion. The purpose of relief venting is to ensure that, in the event of an explosion within the installation, discharge of unburned mixture and combustion products to the atmosphere will occur in time to prevent unacceptably high pressures. The relief vent must be located in such a way that an explosion will not cause harm to personnel when the relief system vents.

There are several ways to achieve efficient relief venting. Bursting discs are widely used. Explosion doors are also widely used, but their effectiveness can be hindered by the inertia of the explosion door. If explosion doors are used, the flow of combustion products and unburned mixture during the relief process may be hindered, and it may be necessary to increase the relief area or increase the mechanical strength of the enclosure to be protected. Vent panels are being used more frequently. These may be blow out panels held in place by special clamps. These panels must be designed so they will release at as low an internal pressure as possible, yet stay in place when subjected to external wind forces. Rupture diaphragm devices may be designed as square, round, or to fit curved surfaces such as silos and cyclones. The use of vent relief devices is discussed in NFPA 68 (1988). Recent developments in venting of deflagrations are also available in NFPA 68 (1988).

Inerting is a very good preventive measure against dust explosions. The maximum oxygen concentration at which dust explosions are "just not possible" cannot be predicted accurately since it depends on the nature of the combustible material; testing is usually required. It has been found that in an atmosphere of 10% oxygen and 90% nitrogen, combustible organic dusts are no longer explosive. To allow a safety margin, it is good industrial practice to maintain oxygen concentrations below 8%. For metal dusts, the allowable oxygen content is ~4% (NFPA 68, 1988). Inert gases containing nitrogen may be supplied by combustion gases from a natural gas burner, nitrogen from a cryogenic unit, nitrogen from a membrane unit, or nitrogen from a pressure-swing adsorption unit. The economics from each of these and other possible sources should be investigated before making a choice. If inerting is used, a reliable, steady source of inert gas must be available. A number of sensors should be installed to be sure the installation can be started only when the oxygen content is below the critical limit and that automatic shutdown happens when this limit is exceeded. Continuous oxygen analyzers, frequently checked for accuracy and operability, should be used. Inerting leads to the possibility of asphyxiation of operating personnel if they are exposed to the inert gas. Strict precautions must be taken to prevent exposure of personnel to inerting atmospheres.

Other inert materials that can be used include water vapor and carbon dioxide, which are somewhat more effective than nitrogen because they have higher heat capacities. To use water vapor effectively as an inerting agent at atmospheric pressure, the temperature should be ~90-95°C. Carbon dioxide is an effective inerting agent, but it has significant solubility in water and other materials and is reactive with alkaline substances.

Explosion suppression systems are designed to prevent the creation of unacceptably high pressure by explosions within enclosures that are not designed to withstand the maximum explosion pressure (NFPA 69, 1986). They can protect process plants against damage and also protect operating personnel in the area. Explosion suppression systems restrict and confine the flames in a very early stage of the explosion, but suppression systems are expensive and require more maintenance than relief venting devices. These may be the reasons this type of safeguard has not been as widely used in industry as one might expect, although its effectiveness has been proven by much practical experience. Explosion suppression systems consist of a sensor system that will detect an incipient explosion and pressurized extinguishers in which rapid action valves will be activated by the sensor. The extinguishing medium is injected into the protected enclosure and evenly dispersed in the shortest possible time. The flames of the explosion are quenched, and the explosion is restricted. There are three types of sensors.

(1) Thermoelectrical—seldom used because they are effective only if installed close to the ignition source, which is usually not known

(2) Optical—which have the drawback of having their action considerably delayed by the dust cloud if a dust explosion occurs; infrared sensors are available that are said to function, even in dust clouds

(3) Pressure sensors—best used for suppression systems since the pressure at an early stage of an explosion in an enclosure will spread evenly and at the

speed of sound in all directions; these detect an incipient explosion with sufficient reliability to activate the extinguisher valves at a predetermined explosion pressure

The extinguishing agent used in dust explosions may be a powder ejected into the enclosure with a gas propellant (usually nitrogen) at 60–120 bar. Effective extinguishing agents include ammonium phosphate and sodium bicarbonate powders, which are popular in Europe. Tests should be made to be sure the extinguishing agent is effective for the dust being used. Occasionally water is used.

In the United States, suppression systems using Halon 1301 (bromotrifluoromethane) (NFPA 69, 1986) to quench the flames in industrial equipment are popular because damage to the product and to electrical components and other equipment is minimized. Halon 1301 is also used for flame suppression in areas occupied by people such as in computer rooms. Extinguishing flames successfully can usually be achieved at Halon 1301 concentrations of $\sim 5\%$ for about 10 min., which usually allows people time to escape the area without harm. Halon 1301 is colorless and odorless and has minimal, if any, central nervous effects to people below a 7% concentration for exposures of ~ 5 min. However, the decomposition products of Halon 1301 that may result from a fire are quite toxic. They have a characteristic sharp, acrid odor that provides a built-in warning system to people. Halon compounds are fluorocarbons which are being phased out of use because of environmental concerns. Substitutes are being developed and should be considered as they become available.

Explosion suppression is a proven technology and should be considered as a candidate for explosion protection. The NFPA has published a standard reference (NFPA 69, 1986) on explosion-suppression protection. Manufacturers should be consulted on design, installation, and maintenance. But, even with explosion suppression, it is common for the explosion pressure to reach 1 atm before it is suppressed. The added pressure surge from the injection of the suppressing agent must also be considered. Therefore, sufficient mechanical strength is always required for enclosures protected by explosion suppression.

G. Substituting Less Hazardous Materials in Processes, Transportation, and Storage

It may be possible to substitute a less hazardous material for a hazardous product. For example, bleaching powder can be used in swimming pools instead of chlorine (Kletz, 1985b). Benzoyl peroxide, an initiator used in polymerization reactions, is available as a paste in water, which makes it much less shock sensitive than the dry form. Phosgene is a highly toxic material used to make N-phenyl carbamate insecticides and other widely used chemicals. It is a key reactant in producing methyl isocyanate (MIC), the highly toxic gas released at Bhopal, India, in December, 1984. Enichem Co., of Italy, has developed a carbamation route to N-methyl carbamate insecticides that uses diphenylcarbonate instead of phosgene. For some uses, PPG Industries is providing diphenyl carbonate powder that is considerably less toxic to ship and handle than phosgene. Other work continues on phosgene substitutes and safer value-added phosgene products. Most of these substitutes involve higher initial costs (Brooks *et al.*, 1986a). Other substitutes have been used to make transportation, storage and processing safer.

(1) Shipping ethylene dibromide instead of bromine
(2) Shipping ethyl benzene instead of ethylene
(3) Storing and shipping chlorinated hydrocarbons instead of chlorine
(4) Storing and shipping methanol instead of liquefied methane
(5) Storing and shipping carbon tetrachloride instead of anhydrous hydrochloric acid; the CCl_4 is burned with supplemental fuel to make HCl on demand at the user's site
(6) Using magnesium hydroxide slurry to control pH instead of concentrated sodium hydroxide solutions, which are corrosive to humans and relatively hazardous to handle; magnesium hydroxide slurry is relatively safe to use in comparison
(7) Using pellets of flammable solids instead of finely divided solids to reduce dust explosion problems

The use of substitutes may appear to be more costly. However, the added safety provided by substitutes may make their use worthwhile and can, in some cases, actually lower the true cost of the project when the overall impact on the process, surrounding areas, and shipping is considered. Substitutes should be employed only if it is known that overall risk will be reduced. Inadequately tested processes and products may introduce unrecognized health, safety, and environmental problems.

H. In Situ Production and Consumption of Hazardous Raw Materials

Some process raw materials are so hazardous to ship and store that it is very desirable to minimize the amount of these materials on hand. Occasionally, it is possible to achieve this by making the hazardous materials on site out of less hazardous materials just before processing so there is only a small amount of the hazardous materials present at any time. An example of *in situ* manufacturing can be found in the batch suspension polymerization of vinyl compounds such as vinyl chloride. One initiator that is useful because of its short half-life and good polymerization properties is diisopropyl peroxydicarbonate (IPP). It has a half-life of 10 hrs as a 50% solution in toluene. This short half-life makes it a desirable initiator. However, storage of IPP is difficult and may be hazardous. A 50% solution of IPP in toluene should be kept between $-20°C$ and $-10°C$. Above $-10°C$, it begins to decompose fairly rapidly, giving off flammable vapors and possibly bursting into flame. Below $-20°C$ it freezes and becomes difficult to handle. A process has been developed (Cox and Shiah, 1970) in which IPP can be made as needed, *in situ*, before use by reacting dilute hydrogen peroxide and isopropyl chloroformate in part of the water phase of the suspension polymerization batch reaction just before polymerization. As a result, there is no need to ship and store IPP. The process is much safer, even though the starting materials (hydrogen peroxide and isopropyl chloroformate) are hazardous chemicals, because IPP is so much more hazardous.

The possibility for *in situ* manufacturing of hazardous chemicals for process use to minimize storage and handling may not be obvious in most processes. It may be appropriate to challenge the chemists, who are best able to develop *in situ* manufacturing, early enough in the process so they can develop the technology necessary to incorporate it into the plant design. These processes should be adequately tested to make certain additional unforeseen risks are not introduced.

I. Using Incineration to Dispose of Hazardous Materials

First consideration should be given to minimizing the production of hazardous waste materials at the source. The next consideration should be to recover, reclaim, and recycle hazardous materials. These activities minimize the need to manage hazardous wastes, resulting in minimal adverse impact on the environment. When all reasonable efforts have been made to minimize the production of hazardous waste materials, and it is impractical or uneconomical to further recover, reclaim, or recycle the remaining hazardous wastes, other means of handling these materials must be considered. Quite often the best solution for the disposal of nonrecyclable hazardous vapors and liquids is to burn them under carefully controlled conditions. The technology of incineration has advanced a great deal, and it is possible to burn hazardous materials routinely and obtain very high levels of destruction removal efficiency of the principal organic hazardous constituents (POHCs). It is common to generate steam with a system designed to burn hazardous vapors and liquids, so it may actually be more profitable and more environmentally sound to burn these materials than to dispose of or treat them in other ways. Chlorinated hydrocarbons may be burned to produce aqueous hydrochloric acid which has a value. The advantage of burning is that it is usually a final solution with no significant emission of hazardous materials. Other methods of recovery usually require some further treatment of the recovered waste (Frey, 1987).

Flares are often used to dispose of flammable gases that are produced at widely varying rates in normal startup and in upset conditions. The discharge of gas to a flare system is inevitably somewhat erratic. In large installations, the flame on a flare stack is often several hundred feet long and can have a heat release rate of 10^7 BTU/hr(Lees, 1980). There is intense radiation from a flare. It is generally necessary, therefore, to have an area around the flare where people may not be located. A large flare can thus sterilize a sizable area of land. The level of heat radiation from a flare that is acceptable at ground level is 1000 BTU/(ft^2)(hr), of which 250 BTU/(ft^2)(hr) is solar radiation. If people are required to work in this area, radiation should be limited to 500 BTU/(ft^2)(hr)(Lees, 1980). If a flare is used, the layout should be designed keeping in mind the large area that may be necessary to suitably isolate the flare. An alternative may be to use enclosed flares. Flare systems can explode because of flashback. Many devices and techniques are available to reduce the possibility of flashback and to prevent air from getting into the system. Open flares should not be used to burn material containing significant quantities of sulfur, chlorine, or other materials that produce toxic compounds when burned. When it is necessary to burn these materials, burners with scrubbing equipment that will remove toxic products of combustion should be used.

Enclosed Flares

Smoke, glare, and noise from a flare are environmentally objectionable. This may inhibit their use, even though from a strict safety point of view, these flares are desirable. Smoke can generally be eliminated or reduced to a low level by methods, often involving the use of steam, which promote combustion. Light and noise from the flame can be eliminated by using enclosed burning systems offered by several companies. John Zink Co. (4401 S. Peoria, Tulsa, OK), for example, offers enclosed flares that can handle up to 200,000 lb/hr of waste gas. The largest of these enclosed flares consists of a field-erected refractory-lined enclosure 44 ft in diameter and 125 ft high. Occasionally they are used along with an elevated flare to handle any waste gas in excess of the enclosed flare's capacity.

Relatively small enclosed flares are often more applicable in the chemical industry. For example, a unit to handle 25 ft^3/min of a waste hydrocarbon gas requires an enclosed, refractory-lined flare about 3 ft in outside diameter and 25 ft in height. Enclosed flare systems have been especially popular in Europe. These enclosed flare systems make it possible to flare waste gas in populated areas or other areas where normal flaring would be objectionable. They should remove some of the reluctance of process designers to use flares, when the flares are a safe way to solve some potential problems.

V. Equipment Design Opportunities

This part will discuss opportunities, primarily in the design of equipment, to make a plant inherently safer. This equipment includes mechanical equipment, piping systems, control systems, and electrical systems. There have been many developments in recent years that process designers can use to improve plant safety and make the plant inherently safer.

A. Avoiding Catastrophic Failure of Engineering Materials

The choices of the construction materials used in a chemical plant are some of the most important decisions a process designer will make. Inherently safer plants will be designed by people who take advantage of the latest and best available technology to avoid catastrophic failure (Liening, 1986a,b). It is not possible for most process designers to be experts in the field of corrosion. Therefore, it is necessary to make use of the resources available.

(1) *Previous plant experience*—this can be an ideal source of information. However, personnel turnover in manufacturing plants may diminish that source of expertise.

(2) *Judgements of professional technical experts*—these people, if they are available and understand corrosion, are often the best source of information, particularly if the problem is new.

(3) *Literature searches*—the problem here is that much of the information available on construction materials is not available in the usual chemical engineering journals and books. Also, there is so much information available, it is hard for the nonspecialist to choose the best information. A technical expert may be helpful here.

(4) *Vendors*—these sources can provide a great deal of information on existing well-known processes but may be of limited help on proprietary or new processes. Also, one must be careful of vendor data, since there is often no way of knowing how valid the information is and obviously, the vendor is usually trying to make a sale.

(5) *Laboratory testing*—sometimes there is no substitute for laboratory tests or tests in existing plant equipment. This can be expensive and time-consuming, but in critical applications, where there is inadequate knowledge, testing is necessary.

1. *Metals*

Uniform corrosion in metals can usually be predicted from lab tests or experience. Corrosion allowances that require thicker metal can be called for in the design of equipment when uniform corrosion rates are expected. However, uniform corrosion is often not the worst thing that can happen to materials. The most important materials failure to avoid in the design of metal equipment is *sudden catastrophic failure*. This occurs when the material fractures under impulse instead of bending. Catastrophic failure can cause complete destruction of piping or equipment and can result in explosions, huge spills, and consequent fires. Some of the more common types of catastrophic failure are described in the following sections.

a. *Low Temperature Brittleness.* Carbon steel becomes more brittle (less ductile) as temperature decreases. Almost any carbon steel will become brittle at $-30°C$ ($-20°F$). Some grades of steel become brittle at surprisingly high temperatures.

(1) A515 steel can become brittle as high as $10°C$ ($50°F$)
(2) A53 steel can become brittle as high as $0°C$ ($32°F$)
(3) Grey cast iron is brittle at all usable temperatures
(4) Silicon cast iron, which is very corrosion resistant, is brittle at all usable temperatures and can fail due to thermal shock

Stainless steels can usually be used safely down to about cryogenic temperatures. Tanker ships carrying liquefied natural gas (LNG) are typically made of steel with 9% nickel added to impart cold temperature ductility properties.

Process designers should check on the possibility of cold temperatures in their process and be sure they have specified the right material for these conditions. Materials used in equipment handling low-temperature fluids should have a ductile–brittle transition temperature below not only the normal operating temperature, but also the minimum temperature that may be expected to occur under abnormal conditions (Lees, 1980).

b. *Stress Corrosion Cracking.* Stress corrosion cracking is a very serious corrosion problem because it has the potential to cause even inherently ductile alloys to fail catastrophically. Many alloys stress corrode crack in a variety of environments, but chloride induced stress corrosion cracking of stainless steel is the most well-known type. This can occur with many grades of stainless steels at 50°C or above. The corrosion depends on temperature, chloride ion concentration, stress, and to some extent, pH. Chloride ion stress corrosion cracking can become worse as a result of residual fabrication stresses, including those from shaping and welding. Stainless steel vessels can corrode and crack from the outside in salty atmospheres, especially near oceans. The corrosion can become worse from insulation that traps salty material on the equipment surface. Corrosion under insulation is a particularly insidious problem because the damage may not be apparent. Typically, corrosion occurs because the weather seal is damaged and moisture enters the insulation. The insulation holds moisture against the metal, possibly concentrating chlorides on the surface, causing continuous attack and extensive damage.

Stress corrosion cracking from chloride ions can usually be avoided by using alloys with a nickel content above $\sim 32\%$. Alloys such as Carpenter 20CB3, Hastelloy G, and Incolloy 825 are high in nickel and are very resistant. Modern testing methods make it possible to predict chloride stress corrosion cracking in the laboratory in a relatively short time with a high degree of confidence.

c. *Hydrogen Embrittlement.* Hydrogen can come from corrosion reactions with water, producing hydrogen that diffuses into the metal. Hydrogen embrittlement occurs most often in high-strength steels such as in relief-valve springs. It can also occur in high-strength stainless steels such as those used in valve stems and springs. Hydrogen sulfide is especially troublesome with high-strength stainless steel.

d. *Other Types of Catastrophic Corrosion.* There are many other types of catastrophic corrosion that can occur, such as

(1) corrosion with very high penetration rates involving pitting and crevice corrosion and galvanic corrosion,
(2) fatigue failure,
(3) creep (usually unlikely under normal conditions but can be caused by misoperation or fire),
(4) mechanical shock which can be caused by water hammer,
(5) thermal fatigue, in which temperature cycles can cause failure of ductile materials
(6) thermal shock, in which high rates of temperature change can cause failure in brittle materials,
(7) zinc embrittlement, in which the amount of molten zinc required to cause embrittlement of stainless steel is small, and at high temperatures, can fail in seconds (Lees, 1980); this type of failure normally occurs only in a fire,

but can also result from welding stainless steel to galvanized steel,

(8) caustic embrittlement, which has been a frequent cause of failure in boilers, and

(9) nitrate stress corrosion, which can result from a high concentration of nitrates, low pH, temperature above 80°C. and high stress.

It was concluded that nitrate stress corrosion caused the crack in the reactor at Flixborough. The reactor was removed and replaced with a 20 in. pipe that failed. The cracking occurred because cooling water containing nitrite had been sprayed on the reactor (Lees, 1980). Many other types of corrosion can occur with catastrophic consequences. The process designer should be aware of special problems that can result in the process and should obtain specialized help if necessary since corrosion is a specialized and very complicated field. It is impossible for most process designers to keep up with corrosion technology without help.

Stainless steel clad over carbon steel can be treated, in terms of corrosion, as a monolithic stainless vessel. This type of construction, which requires special fabrication techniques, is typically used only for heavy wall vessels. A relatively new area of technology that has met with considerable success at Dow is called Resista-Clad.[3] In this process, a thin layer of highly corrosion-resistant (and usually relatively expensive) metal is welded by special techniques. The metal cladding may be several feet wide and is usually welded to a steel (or other metal) substrate, in rows with about 1 in. spacing. The weld bond is about 30% of the total surface. The resulting equipment has the corrosion resistance of the thin layer of cladding and the physical strength of the substrate. It can also withstand a vacuum, which is a big advantage. It can be more tolerant of abuses and upsets, both mechanical and chemical, than other types of less corrosion-resistant materials used in a solid, nonclad form, and still be of reasonable cost compared to less corrosion-resistant materials. It will probably be less likely to fail catastrophically. The metals that can be used for the cladding include tantalum, columbium (niobium), titanium, nickel and some nickel alloys (under development), and zirconium. Thickness can range from 0.02 to 0.1 in. The substrate can include mild steel, stainless steel, aluminum, and copper.

Cost of this process is about 50% more than glass-lined steel and about 67% of equivalent vacuum-proof, all-metal construction of the same metal as the cladding. It can be used in very large vessels and over complex curves. It might be better to use Resista-Clad with a very corrosion-resistant metal than to use a less corrosion-resistant solid metal construction in which there is a significant risk of failure. This may be especially true if there is doubt about the

---

[3] Resista-Clad is a registered trademark of the Pfaudler-USA Co.

resistance to catastrophic failure of the less corrosion-resistant material, or if there is no time to get data.

Explosively clad metals, of which Deta-Clad[4] is an example, are used mostly for small items such as tube sheets and heads for small heat exchangers or vessels. It is usually not economical for large surfaces or complex shapes. Nickel–chromium–molybdenum alloys and some other expensive metals can be applied to carbon steel by explosive bonding.

2. Nonmetals

Glass-lined steel is the material of choice in many applications, particularly in highly acidic conditions and where a smooth, nonsticking, easily cleanable finish is required. The technology has advanced to the point that glass-lined vessels can be made completely free of flaws where corrosion can occur. Agitators and seals can be made that have essentially no exposed metal. Glass linings with improved alkali resistance (up to pH 12) are available, however, glass is not suitable for hydrofluoric acid and hot concentrated phosphoric acid, nor for hot alkaline solutions (*Perry's*, 1984). The chief drawback of glass-lined steel is its brittleness and susceptibility to thermal shock. A nucleated crystalline composite form of glass has superior mechanical properties compared to conventional glass-lined steel. It has three to four times the thermal-shock resistance of glassed steel. However, this form of glass is somewhat less corrosion-resistant, in some conditions, than the best conventional glass-lined steel. Another drawback of glass-lined steel, compared to metals, is its reduced heat transfer capability.

Glass-lined vessels with alloy faced flanges are available at a reasonable cost. Alloy metal flanges on glass-lined vessels should be considered whenever hazardous, flammable, or toxic materials are handled under pressure. The alloy used must be compatible with the glass coating, which must cover the welded joint between the steel shell and the alloy flange. Metal flanges make possible the use of high-integrity gaskets such as spiral wound gaskets, which usually cannot be used on glass surfaces. The metal alloys available for flanges are resistant to a wide variety of extremely corrosive materials.

In general, if the special properties of glass-lined steel are not essential, many manufacturing and process design people prefer to use a highly corrosion-resistant metal instead of glass-lined steel. The brittle nature of glass and its possible failure are of continual concern in any plant that uses glass-lined steel. On the other hand, there are thousands of examples of glass-lined equipment that have served well for many years. Glass failure in a glass-lined vessel does not usually cause catastrophic failure in a short amount of

---

[4] Deta-Clad is a registered trademark of E. I. du Pont de Nemours & Co., Inc., Wilmington, DE.

time because the steel tank itself will usually survive for a significant amount of time. Glass failure may cause holes in the shell to occur, but the vessel does not usually quickly lose its ability to withstand pressure.

3. *Plastic Materials*

The number and combinations of plastics available to the chemical industry is very large and growing. The choices available to the process designer provide many opportunities but also some possible problems. There are numerous advantages in using plastics. They are

(1) lightweight,
(2) good thermal insulators,
(3) good electrical insulators,
(4) relatively easy to fabricate and install,
(5) often cheaper and faster to build,
(6) low in friction factors,
(7) excellent resistors to weak mineral acids,
(8) unaffected by most inorganic salt solutions, and
(9) do not corrode in the electrochemical sense, therefore they are resistant to changes in pH, minor impurities, and oxygen content.

There are also a number of disadvantages in using plastics. They are

(1) limited to relatively moderate temperatures (200°C is high for plastics),
(2) less resistant to mechanical abuse,
(3) high thermal expansion rates,
(4) relatively low strengths (unless reinforced),
(5) often only fair resistors to solvents,
(6) difficult to ground electrically,
(7) poor resistors to organics,
(8) can swell or dissolve in the presence of some solvents,
(9) permeable to some solvents,
(10) flammable and can produce toxic fumes,
(11) much lower heat-transfer coefficients than metals.

Several plastics, with high resistance to chemical attack and high temperatures, deserve special mention for process designers of inherently safer plants. For example, tetrafluoroethylene (TFE), commonly called Teflon brand TFE, is practically unaffected by all alkalies and acids except fluorine and chlorine gas at elevated temperatures, and molten metals. It retains its properties at temperatures up to 260°C. Other plastics that have similarly excellent properties (but are different enough that they each have their niche) include chlorotrifluoroethylene (CTFE); Teflon FEP, a copolymer of tetrafluoroethylene and hexafluoropropylene; polyvinylidene fluoride (PVF_2) (also known as PVDF or Kynar[5]), and perfluoroalkoxy (PFA) (an improved moldable grade of Teflon). These plastics have excellent properties for use in many areas of chemical plants and, when properly used, can provide excellent safety features in preventing leakage, corrosion, and solvent attack. They are not panaceas; each has its limitations, but in the right application, they provide excellent service.

Permeation can be a problem with these materials and can cause corrosion of the substrate. Some of these plastics are significantly less permeable than others. Permeation is especially a problem with Teflon when used with many liquids, including liquids as different as styrene and bromine. Kynar is generally more resistant to permeation than Teflon, but may be more brittle. Cold flow may also be a problem with some of these polymers. The use of glass filled polymers can help overcome cold flow.

Fiberglass reinforced plastics (FRPs) are very popular because they combine the chemical resistant properties of plastics with the considerable strength of glass. Other fibers may be used to reinforce plastics. The lack of homogeneity and the friable nature of FRP composite structures dictate that caution be used in mechanical design, vendor selection, inspection, shipment, installation, and use (*Perry's*, 1984). Many different resins can be used to manufacture FRP items. The choice depends on different mechanical and corrosion-resistance requirements. In general, when hazardous and especially flammable chemicals are involved, FRP vessels are not usually considered to be inherently safe to use for service involving significant pressures, unless great care and extensive testing of similar vessels to destruction is carried out. Large FRP vessels intended for vacuum service, or service where there could accidentally be a vacuum, should be used with great care, if at all. Many FRP tanks have inadvertently been destroyed by being "sucked in" when a vacuum was accidentally applied to them. FRP vessels are also not firesafe, as metal vessels are, and are not usually considered to be satisfactory for use with flammable materials in the chemical industry. It is difficult to construct FRP pressure vessels to meet the conditions of ASME code, Section 10, "Fiber Glass Reinforced Pressure Vessels." Up to this time, it has been required to build two vessels and test one to destruction to get ASME approval.

B. USING ADEQUATE REDUNDANCY OF INSTRUMENT AND CONTROL SYSTEMS

Computer controlled chemical plants are becoming the rule rather than the exception. As a result, it is possible to measure more variables and get more

---

[5] Kynar is registered trademark of the Pennwalt Corporation, Pennwalt Building, Three Parkway, Philadelphia PA 19102.

DESIGN OF INHERENTLY SAFER PLANTS 111

process information than ever before. There is now an opportunity to make chemical plants inherently safer than ever before. However, it must be kept in mind that instruments and control components *will* fail. It is not a question of *if* they will fail, but *when* they will fail, and what the consequences will be. Therefore, the question of redundancy must be thoroughly considered. The system must be designed so that when failure occurs, the plant is still safe. Redundant measurement means obtaining the same process information with two like measurements or two measurements, using different principles.

Redundant measurements can be *calculated* or *inferred* measurements (Grinwis *et al.*, 1986). Two like measurements could be from two pressure transmitters, two temperature measurement, or two level measurements. An example of inferred measurement could be measuring temperature in a boiler and, using a pressure gauge and vapor pressure tables, measuring the pressure to check an actual temperature measurement.

A continuous analog signal that is continuously monitored by a digital computer is generally preferable to a single point or single switch such as a high-level switch or high-pressure switch. A continuous analog measurement can give valuable information about what the value is now and can be used to compute values or compare with other measurements. Analog inputs may be visual and one can see what the set point is and what the actual value is. The software security system determines who changes set points, and it is not easy to defeat.

A single point (digital) signal only determines whether switch contacts are open or not. It can indicate that something has happened, but not that it is going to happen. It cannot provide information to anticipate a problem that may be building up or a history about why the problem happened. Single point signals are easy to defeat. Some single point measurements are necessary, such as fire eyes and backup high-level switches. As a rule, it is best to avoid both pressure transmitters on the same tap, both temperature measuring devices in the same well, both level transmitters on the same tap or equalizing line, and any two measurements installed so that the same problem can cause a loss of both measurements. It is a good idea if possible, to use devices that use different principles to measure the same variable.

An alarm should sound any time redundant inputs disagree. In most cases, the operating personnel will have to decide what to do. In some cases the computer control system will have to decide by itself what to do if redundant inputs disagree. The more hazardous the process, the more it is necessary to use multiple sensors for flow, temperature, pressure, and other variables. Since it must be assumed that all measuring devices will fail, they should fail to an alarm state. If a device fails to a nonalarm condition, there can be serious problems. It is also serious if a device fails to an alarm condition, and there is really not an alarm condition. This is generally not as serious as the first case, but it can provide a false sense of security. Usually it is assumed that two devices measuring the same condition will not fail independently at the same time. If this is assumed, one can consider the effects of different levels of redundancy:

Number of Inputs	Consequence
One	Failure provides no information on whether there is an alarm condition or not.
Two	Failure of one device shows there is a disagreement, but without more information, it cannot be determined whether there is an alarm condition or not. More information is needed; the operator could "vote" if there is time.
Three	Failure of one device leaves two that work; three should be no ambiguity on whether there is an alarm condition or not.

As a matter of interest, if it is assumed that two devices can fail independently at the same time, five separate measuring devices would be required to determine, without ambiguity, that there is an alarm condition, unless more information is available. The vote would be "three out of five" as the consensus to determine if an alarm condition existed. This is rarely done except in the nuclear and aerospace industries.

Large continuous processes exist that are difficult to shut down, and a shutdown may be almost as dangerous or more dangerous than continuing to run in an alarm condition. When these processes do shut down, they must shut down in an orderly manner to avoid damage and hazards. If there is no time to wait for an operator to "vote," or if it is necessary for the operator to have more information, the computer must make the decision, and three or more measuring methods should be considered. The control mechanism should be "intelligent" enough to detect a problem and correct it before the problem gets to a serious stage. This may require corrective action at an early stage. There are numerous examples of this.

(1) If a pump starts up, downstream pressure should go up within a certain time interval. If it doesn't, there should be an alarm.

(2) Materials that react exothermically should require cooling when they are continuously added to a reactor. If cooling is not being called for, the materials are probably not reacting as they should, or there may be an error in flow measurements and there should be an alarm. If the reactants build up in the reactor without reacting as they should, there could be a "bomb" of reactants building up in the process, and waiting to react.

(3) If a signal, such as a level indicator signal, is usually noisy, and it becomes "frozen" on one value, the line to the sensor may be plugged. The computer system should detect this and give an alarm.

(4) The computer system should detect a value that is not following an expected trend. An example is a level indicator or weight indicator on a tank

that is being filled using a pump. If the weight does not increase as the pump operates, something is wrong and there should be an alarm.

(5) If a tank is being filled from another tank, the level or weight on the tank being filled should go up by the same amount as the tank being used to fill it goes down. The computer should check this and give an alarm if it doesn't happen.

(6) A device, such as a temperature or pressure measuring device, that suddenly shows a full-scale or down-scale reading has probably failed. The computer system should detect this and give an alarm.

In a computerized plant, it is best to avoid putting redundant inputs on or in the same slot, cluster, or board in the computer. The safety system should be continuously exercised. An independent safety system suffers from a major drawback, particularly if it is implemented using relays. This drawback is that the safety system usually remains inert for long periods of time. Electronic circuits or relays are subject to failures such that when the safety system is called on to operate, it may be incapable of doing so. The best way to ensure that the safety system has not suffered a fault is to continuously exercise it. This is best done if the safety system is also responsible for control calculations and is constantly in use (Wensley, 1986).

1. Critical Instrument System

Critical instrument systems are any instrument systems that must function properly for the safe operation of the process or to protect equipment. A major concern is instrumentation that has the potential for shutting down a process or unit operation. In computer-controlled plants, not only are the field devices critical. The system that reads the signal, interprets it, and displays it is also critical. Critical instrument systems must be tested and maintained regularly, and recalibrated if necessary. The instrument systems that are associated with the critical instrument should be tested, not just the instrument itself. An "insult" test is preferred—one that simulates a condition the instrument should detect, such as a high level alarm or high oxygen alarm. The insult test should activate the entire system from the sensor to the activating device it is supposed to activate. The design of the plant should take the need for testing into account and make it as easy as possible to test critical instruments systems regularly.

2. Fail-Safe Valves

Valves can have four modes of operation during failure: Fail open, fall closed, fail in the current position, and fail in any position without having an impact in the process (this is rarely a valid option). In order to fail safe, most valves should usually fail closed. There are exceptions. Often cooling water valves and valves in coolant circulating systems should fail open. Diverter valves in solids-handling systems should usually fail in the current position. There are very few cases where it does not matter how a valve should fail.

Many of the fail-safe valves used in chemical processes are quarter-turn valves. Generally, ball valves with spring loaded actuators are recommended so that there is positive fail-safe movement in case of power or air failure. It may not be practical for very large valves to be spring loaded. These large valves should generally have a local air supply tank to cause them to fail to the fail-safe condition.

C. Properly Designing Pressure Relief Systems

Emergency pressure relief systems should be installed to protect vessels and other equipment and their surroundings from the dramatic, often catastrophic, effects of an overpressure and subsequent failure. The process designer of inherently safer plants should use the best tools available to design appropriate pressure relief systems. Fortunately, since 1984 there have been significant developments in the complex field of pressure relief design that now make possible realistic system design. This pressure relief discussion differs from what is commonly referred to as "explosion venting." Events such as dust explosions and flammable vapor deflagrations propagate nonuniformly from a point of initiation, generating pressure and shock waves. Such venting problems are outside the scope of this discussion of pressure relief systems.

The design of relief systems involves, in general, the following steps:

(1) Generate a scenario—what could reasonably happen that could cause high pressures? This could be fire, runaway reactions, phase changes, or leaks from high pressure sources.

(2) Calculate the duty requirements, the lb/hr of material that has to be vented and its physical conditions including temperature, pressure, ratio of vapor to liquid, and physical properties. This is a rather involved calculation.

(3) Calculate the area required to relieve the pressure, based on the duty, inlet and outlet piping, and downstream equipment. This is also a rather involved calculation.

(4) Choose the pressures relief device to be used, which should be specified from vendor information.

A group of chemical companies joined together in 1976 to investigate emergency relief systems. This later resulted in the formation of The Design Institute for Emergency Relief Systems (DIERS), a consortium of 29 companies under the auspices of the American Institute of Chemical Engineers. DIERS was funded with $1.6 million to test existing methods for emergency relief system designs and to "fill in the gaps" in technology in this area, especially in the design of emergency relief systems to handle runaway reactions (Fisher, 1985). DIERS completed contract work and disbanded in 1984.

Huff was the first to publish details of a comprehensive two-phase flow computational method for sizing emergency relief devices, which with refinements, has been in use since 1977 (Huff, 1973, 1977, 1984b). The most significant theoretical and experimental finding of the DIERS program is the ease with which two-phase vapor-liquid flow can occur during an emergency relief situation. The occurrence of two-phase flow during runaway reaction relief almost *always* requires a larger relief system compared to single-phase vapor venting. The required area for two-phase flow venting can be from about twice the area to much more than than this in order to provide adequate relief than if vapor-only venting occurs (Huff, 1977). Failure to recognize this can result in drastically undersized relief systems that will not provide the intended protection.

Two-phase vapor-liquid flow of the type that can affect relief system design occurs as a result of vaporization and gas generation during a runaway reaction. Boiling can take place throughout the entire volume of liquid, not just at the surface. Trapped bubbles, retareded by viscosity and the nature of the fluid, reduce the effective density of the fluid and cause the liquid surface to be raised. When it reaches the height of the relief device, two-phase flow results.

An area that has received considerable attention is the downstream and upstream piping and equipment around a relief device. It is realized that these piping systems can be the flow-limiting elements for the entire relief system. The entire piping system, in which the relief system is located, as well as the relief device itself, must be considered. The expertise for designing relief systems for complex systems has been developed to a high degree. However, many systems are complicated, and the knowledge of exactly how to use the calculational systems described in the literature is not widely available.

The *DIERS Project Manual* (published by the American Institute of Chemical Engineers) is a helpful compendium for experienced safety-relief system engineers. Extensive background and experience are required to properly understand and apply the methodology. Help is available from the DIERS contractors and the DIERS Users Group.

DIERS sponsored development of a comprehensive computer program that can be used to size emergency relief systems for runaway reactions in industrial vessels, if appropriate information is available. This program has considerable potential for widespread use throughout the chemical industry (Grolmes and Lung, 1985). The JAYCOR Corp. has modified an existing proprietary program which is also capable of emergency relief design calculations (Klein, 1986).

An accelerating rate calorimeter (ARC) can be used to provide design values for emergency pressure-relief flow requirements of runaway systems. The ARC is a device used to obtain runaway history of chemical reactions in a closed system (DeHaven, 1983; Huff, 1982, 1984a; Townsend and Tou, 1980). The experimental technique is fairly straightforward, but considerable engineering expertise is required to do the calculations needed to design a relief system from the ARC data.

Another very useful tool called a "bench scale apparatus for characterizing runaway chemical reactions" or Vent Sizing Package (VSP) has been developed and can handle largely unknown systems with a small test cell ($\sim$ 100 ml) compared to 9 ml for the ARC. This is an extension of ARC technology, but less engineering expertise is required to do the actual vent sizing. This device allows direct and safe extrapolations on a large scale at relatively low cost. Vents may be sized with less information than is required with other methods. Key physical properties should be known for the processing materials involved. They can be either measured or estimated, although it is not necessary to know the identities of the chemicals involved. Before using the apparatus, differential thermal analysis (DTA) data, or equivalent data, on the materials to be used should be available. This can help determine if the system has such rapid energy release possibilities that detonations or rapid releases of energy could occur, which would be unsafe in the 100 ml test cell. Given the upset condition, the new method allows for the safe extrapolation to full-size process vessels (Fauske and Leung, 1985).

The VSP makes sizing emergency relief systems possible without a comprehensive computer program. The device is a bench scale experimental apparatus that will measure runaway reaction data under adiabatic conditions in a vessel with very low thermal inertia, or phi (ϕ) factor (phi denotes thermal inertia, which is the ratio of the heat capacity of the sample plus the bomb to the heat capacity of the sample). With this equipment, it is possible to predict the vapor-liquid disengagement regime and viscous vs. turbulent flow-pipe behavior. Useful data can be obtained as high as 400 psig. It has a typical thermal inertia, or ϕ factor, of 1.05. The equipment can withstand up to 2500 psig and is hydrostatically tested to 7500 psig.

The ability of the VSP to characterize the vapor-liquid disengagement regime and viscosity effects is important because of the dramatic effect these variables have on relief system size (Fauske and Leung, 1985). No other apparatus is known that can predict these variables under runaway conditions. Figure 6 shows schematically how the unit is designed.

The selection of an actual pressure relief valve to use when the area is known is not a trivial problem, since there are many vendor catalogs and each can be

DESIGN OF INHERENTLY SAFER PLANTS 117

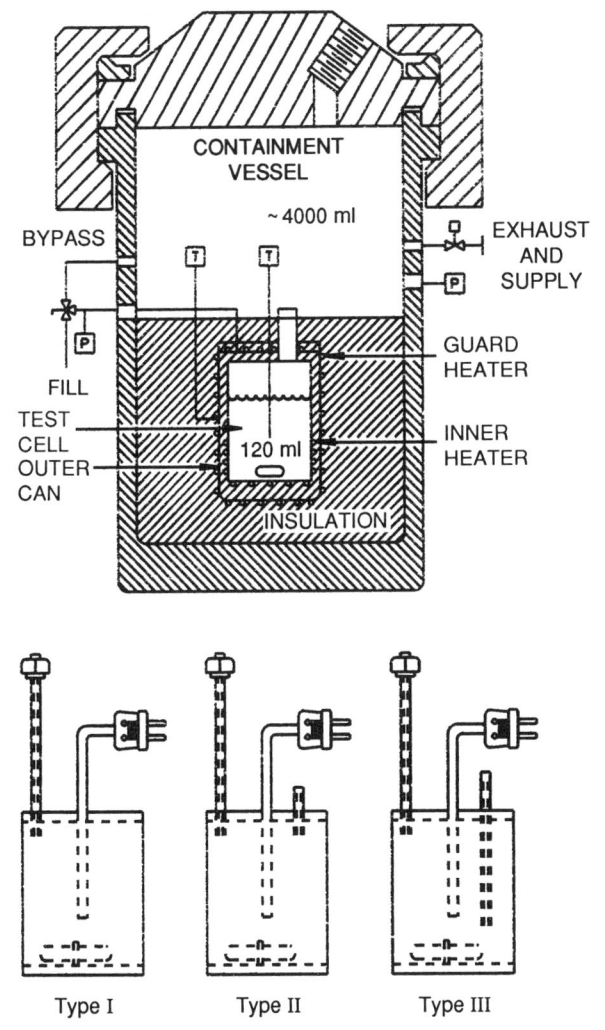

FIG. 6. Small-scale test equipment with closed and open test cell designs. Type I test cell, closed system thermal data; type II test cell, open system vent sizing and flow regime data; type III test cell, open system viscous effects data.

fairly difficult to use unless one is experienced with it. A selection program using a personal computer is available which makes it easy to select the model number for several manufacturers of relief devices (flanged full-nozzle valves only). It will also calculate the API and ASME code areas for certain design cases, but does not consider all cases. It provides good documentation. This program is called SARVAL[6].

When using safety valves to relieve pressure vessels, consider isolating them from the process by using rupture disks (frangible safeties). There are several reasons for this.

(1) They will prevent the process components from leaking into the atmosphere, which can happen with relief valves. Conventional relief valves have an allowable leakage rate; rupture disks do not.

(2) Relief valve life is extended since process components will not contact the valve.

(3) They can possibly extend the time between valve overhauls.

(4) Less-expensive valve material can be used.

Space between a rupture disk and a safety relief valve must be provided along with a monitoring system, or other indicator, to detect disk rupture or leakage. This space should also be provided with a small relief valve to prevent pressure from building up between the relief valve and rupture disk, possibly causing the system to effectively have a higher bursting pressure than intended.

Chemical process fluids can be too hazardous to permit direct venting to the atmosphere. In this case, the vent system can take on the character of a separate chemical processing unit. The configuration of such a vent system is shown in Fig. 7 (Huff, 1987). In this system, provision is made to handle an appreciable amount of liquid that may vent, along with the gases and vapors, from the uncontrolled reaction. This liquid is separated from the gas-phase materials in either a large catch tank or a cyclonic-type separator. The vapors and gases are then treated in a unit such as a scrubber. The discharge from the scrubber can be situated in order to assure adequate plume dispersion in the event of misoperation of the scrubber. If desired, the discharge could go to a flare or to an incinerator equipped with a scrubber to remove toxic compounds formed by burning, such as hydrochloric acid.

D. PROVIDING SAFE AND RAPID ISOLATION OF PIPING SYSTEMS OR EQUIPMENT

It should be possible to easily isolate fluids in equipment and piping when potentially dangerous situations occur. This can be done using emergency block valves (EBVs). An EBV is a manual or remotely-actuated protective device that should be used to provide manual or remote shut-off of

---

[6] SARVAL is available from Kenonics Controls, 3667 60th Avenue SE, Calgary, Alberta T2C 2E5, Canada.

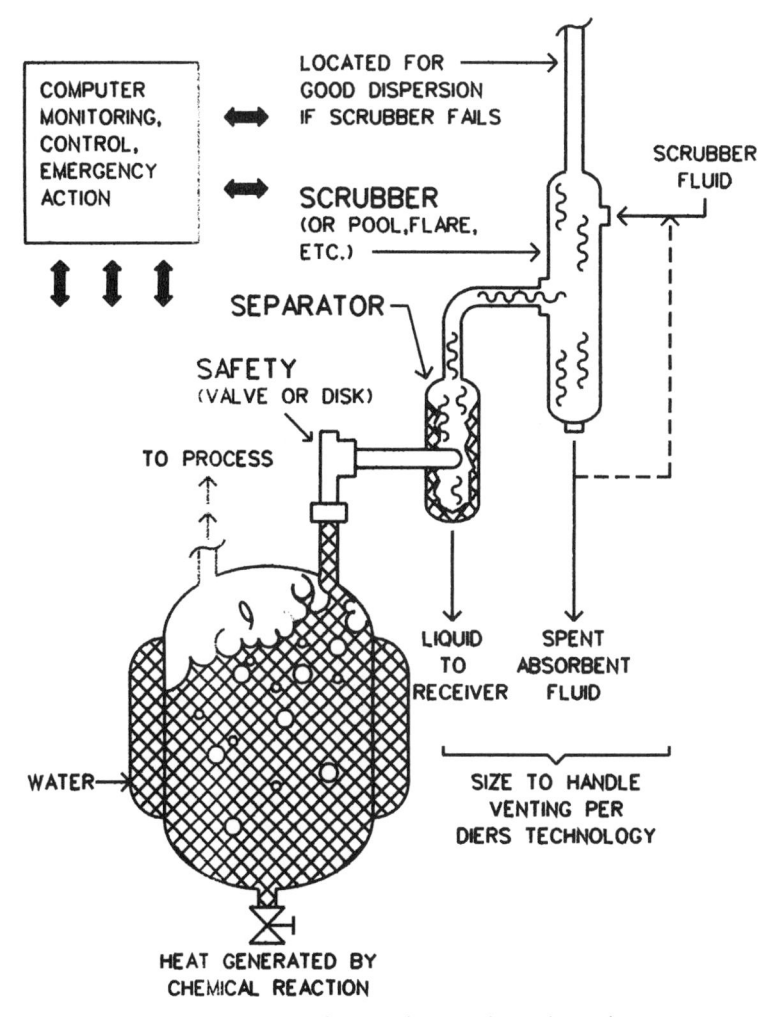

FIG. 7. Typical system for two-phase venting and containment.

uncontrolled gas or liquid flow releases. EBVs can be used to isolate a vessel or other equipment, or an entire unit operation. Manual valves are often used on piping at block limits where it is unlikely there would be a hazard to personnel if an accident occurs. Remotely controlled EBVs are recommended on tanks and on piping in areas where it may be hazardous for personnel in case of an accident or where quick response may be necessary. EBVs used on tanks should be as close as possible to the tank flange and not in the piping away

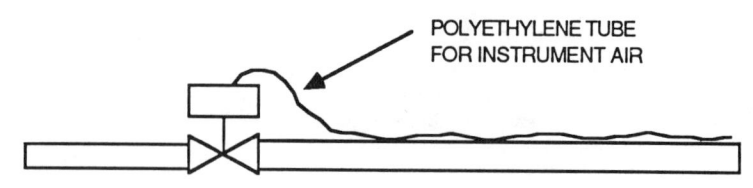

FIG. 8. Polyethylene tubing on fail-safe, spring-loaded, fire-safe valve used as an emergency block valve.

from the tank. In cases where EBVs may be exposed to fire, the valve and valve operator must be fire safe. The valve actuators for remotely controlled EBVs should be air, nitrogen, or hydraulic-pressure operated, with a spring to close to the fail-safe position. The air or nitrogen connection can be polyethylene tubing which will act as a fusible link, causing the valve to close if the tubing is melted or damaged resulting in a loss of air pressure (see Fig. 8). An alternative is to use heat actuated (such as a fusible link) valves. This could be in conjunction with separately operated remote valves, if desired.

In some cases, after careful evaluation, other valves may be considered for EBVs such as spring-loaded control valves that fail closed, back flow check valves (these are not normally considered reliable enough for EBVs by many engineers), and excess flow valves. Excess flow is the loss of material from the confined environment of a vessel or pipeline. Two approaches are available for the detection and valve action of excess flow valve systems: (1) External, where excess flow is detected outside the valve itself, and (2) internal, which is within the valve unit and has limited applications. Excess flow conditions are detected more readily because of loss of resistance to flow than because of loss of pressure. All excess flow detection systems are based on product physical properties as well as flow rate. A change of products or process conditions may require a change in the excess flow detection system. For example, a number of excess flow valves are rated for low pressure and are made of materials not suitable for hydrocarbons.

E. Using Piping, Gaskets, and Valves that Take Advantage of Modern Technology (Jackson, 1986)

1. *Piping*

All-welded pipes and flanges should be used in the inherently safer chemical plant. Since flanges are a potential source of leaks, as few flanges as possible should be used. This, of course, has to be realistic. If it is necessary to clean out pipes, flanges must be provided at appropriate places to make cleaning possible. Also, enough flanges must be provided to make maintenance and installation of new equipment reasonably easy. Screwed piping should be avoided for toxic and flammable materials. It is very difficult to make screwed fittings leakproof, especially with alloys such as stainless steel. Where screwed piping is necessary, use Schedule 80 pipe as a minimum. Pipe nipples should never be less than Schedule 80.

Pipe support design should be given special attention. It may be desirable to increase pipe diameter to provide more pipe strength and rigidity and make it possible to have greater distance between supports. Normally in chemical plants, it is not desirable to use piping less than $\frac{1}{2}$ in. in diameter and preferably not less than 1 in. in diameter, even if the flow requirements permit a smaller pipe, except for special cases. Pipe smaller than $\frac{1}{2}$ in. has insufficient strength and rigidity to be supported at reasonable intervals. Tubing should normally be used for anything smaller than $\frac{1}{2}$ in. Tubing is not as fragile as pipe in small sizes. It can be bent, which reduces the number of fittings required. If it is necessary to use smaller pipe or small tubing, special provisions should be made for its support and mechanical protection. Also, consideration should be given to using schedule 80 or schedule 160 pipe if small pipe is required to provide extra mechanical strength, even if the fluid pressure does not require it.

2. *Gaskets*

Gaskets are among the weakest elements of most chemical plants. Blown out or leaky gaskets have been implicated in many serious incidents. A leak at a flange can have a torch effect if ignited. A fire of this type was considered as a possible cause of the Flixborough disaster (Lees, 1980). Modern technology makes it possible to greatly reduce the incidence of gasket failure by using spiral wound gaskets. These are sold by several manufacturers, including Flexitallic, Parker Spirotallic, Garlock, and Lamons. When used properly, spiral wound gaskets are usually safer to use than most other types of gaskets. The preferred spiral wound gasket has an inner and outer metal ring which makes it virtually impossible for chunks of the gasket to blow out. The outer ring, called a gauge ring, makes it possible to accurately line up the gasket between the flange bolt holes. When the gasket is tightened, it is tightened down to the gauge ring which automatically provides the proper compression. The inner ring protects the spirals from spreading into the interior of the pipe and makes proper compression possible. Typical spiral wound gaskets are shown in Fig. 9.

Spiral wound gaskets cost about four to eight times more than compressed asbestos gaskets, but are easily worth it if they can prevent blowouts and leaks. Also, they generally last considerably longer than compressed asbestos-type gaskets. Spiral wound gaskets can be made in virtually any metal and filler

FIG. 9. Spiral wound gaskets.

combination. In many cases, the inner ring, which contacts the process fluid, can be stainless steel, and the outer ring can be steel. Typically the filler material is Teflon, Kevlar (an aramide fiber made by du Pont), or graphite, and the metal spirals are stainless steel. Graphite spiral wound gaskets are fairly expensive and fragile, but they are very resistant to chemicals and high temperatures.

When using 150 lb flanges with spiral wound gaskets, use only weld neck or lap-joint type flanges. Avoid the use of slip-on or threaded flanges because they are not strong enough and their use is discouraged by the American National Standards Institute (B16.5.). It is important that fire resistive gaskets be used with fire-safe emergency block valves.

Bolting with spiral wound gaskets is very important. Plain carbon steel bolts, such as A307 Grade B, should never be used with spiral wound gaskets. They are not strong enough. High-strength alloy bolts such as A193-B7, which contains Cr and Mo, should be used with A194 heavy hex nuts. To properly seal spiral wound gaskets, it is necessary to tighten the bolts to specified torque limits, which are generally higher than with conventional gaskets.

Spiral wound gaskets are not a solution to all gasket problems. They are usually not satisfactory for use with strong acids unless the metal exposed to the acid can tolerate it. Often Teflon envelope gaskets are better for such applications. Use the milled type or U-type Teflon envelope gaskets. Avoid the use of slit-type Teflon envelope gaskets. Spiral wound gaskets cannot be used on vessels with glass-lined flanges. Teflon envelope gaskets and some other types of gaskets are usually better on glass-lined surfaces. Teflon envelope gaskets can burn out in a fire and in some cases can be blown out. However, it is possible to specify metal flanges made of highly corrosion-resistant metal on glass-lined vessels. If this is done, spiral wound gaskets can be used on glass-lined equipment. A glass-lined reactor with metal flanges would usually be considered inherently safer than a glass-lined reactor with glass-lined flanges, assuming the metal used on the flanges can withstand the corrosion of the system.

3. *Valves*

It is desirable and inherently safer to use fire-safe valves whenever it is necessary to isolate flammable or combustible fluids in a pipeline, tank, or other type of equipment. Fire-safe valves should be considered for handling most fluids that are highly flammable, highly toxic, or highly corrosive and that cannot be allowed to escape into the environment. Fire-safe valves should also be used to isolate reactors, storage vessels, and pipelines. They can be used wherever EBVs are required.

A fire-safe ball valve is a valve that is free to move slightly, with pressure in the line, in order to contact a secondary metal seat if the line has been heated enough to melt the plastic seats (usually made of Teflon, but can be made of other plastics) used in the valve. The ball valve has fire-safe stem and body seals. Both ball valves and high-performance butterfly valves can be made fire-safe. Sleeve-type plugcocks are not normally considered completely fire-safe because their construction does not allow the plug to move if the plastic sleeve is melted. Lubricated plug cocks, globe valves, and gate valves are fire-safe if they are built with metal-to-metal seats and asbestos or graphite packing. However, these valves have very limited use in the chemical industry. It has been found that lubricated plug-cocks usually become inoperable if not cared for properly. The stem and body corrode and stick together if they are not lubricated and operated regularly. Lubricated plug-cocks should generally be avoided in the chemical industry.

With the increase in popularity of automated plants, quarter-turn valves are very popular and are used in most installations. The only common quarter-turn valves that are available as completely fire-safe valves are ball valves and high-performance butterfly valves. There are other special fire-safe valves, that are not common, which are used for special purposes. When fire-safe butterfly valves are required, they should pass the Exxon BP-3-14-4 (Modified) Fire Test. The basis for fire-safe valves in general is the API 607 standard Test for Firesafe Soft-Seated Valves.

4. Dry Quick-Disconnect Couplings

Spillage from regular and accidental disconnections of fluid couplings used at liquid-transfer points can be reduced by the use of dry quick-disconnect couplings. These devices combine a coupling connection that is easy to connect with a built-in valve that automatically closes unless the coupling is connected. This can minimize the hazards of handling toxic, flammable and corrosive liquids. They are especially useful for tank trucks and portable tanks where frequent connecting and disconnecting of the coupling is required. One type of high-quality dry quick-disconnect coupling is the Kamvalok coupling*.

5. Spring-Loaded Check Valves

When check valves are required, spring-loaded check valves provide more positive shut off than swing check valves. Swing check valves depend on gravity, and won't work if improperly installed. Occasionally, they won't work well even if they are properly installed. Swing check valves must be installed either horizontally or "vertically up." It is very easy to install them in the wrong position. Chances that a check valve will work are better if it is spring loaded than if it is a swing type. Positive shut off is available if Teflon seats can be used. If a small leak can be tolerated, metal-to-metal contact can be specified. It has been found that the potential for serious water hammer is much reduced if spring loaded check valves are used instead of swing checks. Swing checks slam shut at the last instant, possibly causing serious water hammer. Spring loaded check valves close smoothly and slowly, reducing the possibility of serious water hammer.

6. Plastic Pipe and Plastic Lined Pipe

Plastic lined pipe is excellent for many uses such as highly corrosive applications, where sticking is a problem, and where ease of cleaning is a factor. It is often the cheapest alternative. However, if a fire occurs there may be "instant holes" at each flange because the plastic will melt away, leaving a

---

* Manufactured by the Fluid Handling Group of the Dover Corp., Cincinnati, OH.

gap. Therefore, plastic lined pipe should not ordinarily be used for flammable materials that must be contained in case of a fire. An exception to this is a firesafe plastic lined pipe system made by the Resistoflex Corp., which provides a metal ring between each flange that will make plastic lined pipe firesafe. The pipe will probably have to be replaced after a fire, but the contents of the pipe will be contained during a fire.

In general, all types of solid plastic or glass-reinforced plastic pipe should not be used, if possible, with flammable liquids. Compared to metal, plastic piping melts and burns easier, is more fragile, is easily mechanically damaged, is harder to adequately support, and should be used with appropriate judgement.

F. USING STRONG VESSELS TO WITHSTAND MAXIMUM PRESSURE OF PROCESS UPSETS

It is sometimes possible to anticipate the worst reasonable process upset and design the process to withstand these conditions without relieving the contents through a pressure relief system. For example, it is possible to carry out a simulated styrene runaway polymerization reaction in an ARC apparatus (Section IV,A) to determine the highest pressure and temperature the system can achieve. Under certain conditions the actual composition used in the plant will contain some solvent and polymer. It may turn out, for example, that the maximum pressure reached by adiabatic polymerization is ~ 300 psig when the reaction starts at the normal operating temperature of 120°C. With this knowledge, it is possible to design polymerization equipment that will withstand this pressure, plus a reasonable safety factor, with the assurance that a runaway reaction will not cause a release of material or equipment rupture. The extra cost of the high pressure system may be justified not only by the extra safety. With high pressure systems, it may also be unnecessary to have elaborate collection equipment for the polymer and volatile material that may be released during runaways that cause venting from the pressure relief system from lower pressure reaction equipment.

It is still necessary to have a small relief system to allow for thermal expansion of a liquid-full system. This relief system is also necessary for handling hydraulic overfill and fire conditions, but the system is usually relatively simple.

Deflagration pressure containment is a technique for specifying the design pressure of a vessel and its appurtenances so that they are capable of withstanding the pressure that results from an internal deflagration. This may be inherently safer than relying on techniques to prevent deflagrations. These techniques are not to be used to contain a detonation. The ASME Boiler and

Presssure Code, Section VIII, provides guidelines for designing deflagration pressure containment. The design pressure is based either on preventing rupture of the vessel (i.e., on the ultimate strength of the vessel) or on preventing permanent deformation of the vessel (i.e., on the yield strength of the vessel) from internal positive overpressure. Because of the vacuum that can follow a deflagration, all vessels, in which deflagration pressure containment design is based on preventing deformation, shall also be designed to withstand a full vacuum (NFPA 69, 1986).

The possibility of equipment, which normally runs at or near atmospheric pressure, going into a vacuum condition should also be considered. Vacuum relief systems and vacuum breakers don't always work. On a hot summer day, a sudden rainstorm can cool a large tank or hopper very rapidly. This can cause rapid cooling and contraction of air inside the tank and condensation of vapors, if they are present. This can cause very rapid lowering of pressure inside the tank, which can implode the tank if insufficient provisions have been made for air to enter the tank. These conditions are common in northern climates where the rate at which a vacuum can be produced may be increased in cold weather. In many instances, vacuum devices that are supposed to work are frozen and inoperative and the tank implodes. It may be inherently safer to design tanks to handle a vacuum than to depend on vacuum relief devices alone.

G. Avoid Inherently Unsafe Equipment

Some equipment items are regarded as inherently unsafe for use in flammable or toxic service and should be avoided if possible. The items included are described in the following sections.

1. Glass and Transparent Devices

Glass devices such as sight glasses, bulls eyes, sightports, rotameters, and glass and transparent plastic piping and fittings are sensitive to heat and shock. Transparent plastic devices may be resistant to shock, but are not resistant to high temperatures (*Loss Prevention Principles*, 1986). If they fail in hazardous service, severe property damage and personnel injury can result. Two guidelines to consider are (1) if broken, would they release flammable material, and (2) if broken, would they expose personnel to toxic or corrosive materials? Some suggested ways to avoid glass and transparent devices are listed below.

Glass item	Nonglass equivalent
Glass rotameter	Magnetic flowmeter or dP cell
Bubbler	Pressure switch
Level gauge	Capacitance probe, conductivity cell, float gauge, magnetic liquid level detector, nuclear level detector
Glass or plastic pipe or tubing	Glass-lined steel pipe or plastic lined steel pipe
Sight port	Appropriate instrumentation or fiber optics with a video camera

If, after careful review, it is necessary to use a transparent device that could result in a hazard if broken, design safeguards should be used such as (1) shields or covers, (2) extra-strong devices (such as 300 psig equipment in 100 psig service), or (3) suitable excess-flow valves or remotely operated isolating valves.

2. *Flexible or Expansion Joints*

Eliminate flexible or expansion joints in piping wherever possible. Flexible joints and expansion joints are any corrugated or flexible transition devices designed to minimize or isolate the effects of thermal expansion, vibration, differential setting, misalignment, pumping surges, wear, load stresses, or other unusual conditions. *Almost without exception*, when a flexible joint is installed in a piping system, the flexible joint becomes the weak link in the system (*Safety Standards*, 1982). If, after considering all reasonable alternatives, it is necessary to use a flexible joint, make certain the temperature and pressure rating of the flexible joint are adequate for these conditions. The flexible joint system must be protected from overpressure, and provisions for isolating the flexible joint system should be provided. The need for flexible joints can sometimes be eliminated by properly designing piping so that solid piping will be able to handle misalignment and thermal changes by bending slightly. This is generally much preferred to using an expansion joint, which may be a weak point in the system.

It has been found at Dow, for example, that in many cases, electronic load cells can be used, with no flexible or expansion joints required, to accurately weigh large reactors or process tanks that have many pipes attached to them. This is done by "cantilevering" the pipes attached to the reactor or tank, and using sufficient runs of straight horizontal unsupported piping to take up movements and vibration without interfering significantly with the operation of the load cells. Flexible joints should not be used as a correction for piping errors.

H. Using Pumps Suitable for Hazardous Service

A wide variety of excellent pumps is available in the chemical industry. It is sometimes a problem to choose the best from the large number available. This discussion will be limited to centrifugal pumps. Assuming that one has sized the pump, decided on a centrifugal pump, and has chosen a suitable list of vendors, the remaining main choices to make are (1) the metallurgy to be used, (2) whether to use seal-less pumps or conventional centrifugal pumps, and (3) what type of seal, if using conventional pumps (Cromie, 1986).

1. *Metallurgy*

Don't use cast iron for flammable or hazardous service unless there is no other choice. In very few cases, only cast iron is available. This is true for large double-suction water pumps used by municipalities where types other than cast iron are not readily available. The minimum metallurgy for centrifugal pumps for hazardous or flammable materials is cast ductile iron, type ASTM A 395, having an ultimate tensile strength of $\sim 60,000$ psi. This metal is not brittle at ordinary temperatures. Often, special alloys are required because of corrosion. In no case should brittle materials be used for a pump if other choices are available.

2. *Seal-less Pumps (Reynolds, 1989; Cromie, 1986)*

Seal-less pumps are becoming very popular and are widely used in the chemical industry. Mechanical seal problems account for most of the pump repairs in a chemical plant, with bearing failures a distant second. The absence of an external motor and a seal is appealing to those experienced with mechanical seal pumps. However, do not assume that just because there is no seal, seal-less pumps are always safer than pumps with seals, even with the advanced technology now available in seal-less pumps. Use seal-less pumps with considerable caution when handling hazardous or flammable liquids.

Seal-less pumps are manufactured in two basic types: canned motor and magnetic drive. Magnetic drive pumps have thicker "cans" which hold in the process fluid, and the clearances between the internal rotor and "can" are greater compared to canned motor pumps. This permits more bearing wear before the rotor starts wearing through the "can." Because most magnetic drive pumps use permanent magnets for both the internal and external rotors, there is less heat to the pumped fluid than with canned motor pumps. With magnetic drive pumps, containment of leakage through the "can" to the outer shell can be a problem. Even though the shell may be thick and capable of holding high pressures, there is often an elastomeric seal on the outer magnetic rotor with little pressure capability.

Canned motor pumps typically have a clearance between the rotor and the containment shell or "can," which separates the fluid from the stator, of only 0.008 to 0.010 in. The "can," which is typically 0.010–0.015 in. thick and made of Hastelloy, has to be thin to allow magnetic flux to flow to the rotor. The rotor can wear through the "can" very rapidly if the rotor bearing wears enough to cause the rotor to move slightly and begin to rub against the "can." The "can" may rupture causing uncontrollable loss of fluid being pumped. Some canned motor pumps have fully pressure-rated outer shells which enclose the canned motor; others don't.

Both canned motor and magnetic drive pumps rely on the process fluid fo lubricate the bearings. If the wear rate of the bearings in the fluid being handled is not known, the bearings can wear unexpectedly, causing rupture of the "can." Running a seal-less pump dry can cause complete failure. If there is cavitation in the pump, hydraulic balancing in the pump no longer functions and excessive wear can occur leading to failure of the "can."

A number of liquids require special attention when applying canned motor and magnetic drive pumps. For example, a low-boiling liquid may flash and vapor-bind the pump. Some liquids with high specific gravities (above about 1.3–1.7 cP) can cause the rotor to become magnetically uncoupled from the stator, which can cause considerable heat to be generated. Solids in the liquid can also be bad because of close clearances in the pump. Low viscosity (in the range of 1 to 5 cP) fluids are normally poor lubricators and one should be concerned about selecting the right bearings. For viscosities below 1 cP, it is even more important to choose the right bearing material.

A monitor to detect bearing wear is available on some seal-less pumps but they generally don't offer complete monitoring of all internal bearings for axial and radial wear. Seal-less pumps typically run so smoothly and quietly that it is usually not possible to determine by vibration or noise if a bearing is badly worn.

A mistreated seal-less pump can rupture with potentially serious results. The "can" can fail if valves on both sides of the pump are closed and the fluid in the pump expands either due to heating up from a cold condition, or if the pump is started up. If the pump is run dry for even a short time, the bearings can be ruined. The pump can heat up and be damaged if there is insufficient flow to take away heat from the windings. Seal-less pumps, especially canned motor pumps, produce a significant amount of heat since nearly all the electrical energy lost in the system is absorbed by the fluid being pumped. *If this heat cannot be properly dissipated, the fluid will heat up with possibly severe consequences.* Considerable care must be used when installing and maintaining a seal-less pump to be sure that misoperations cannot occur.

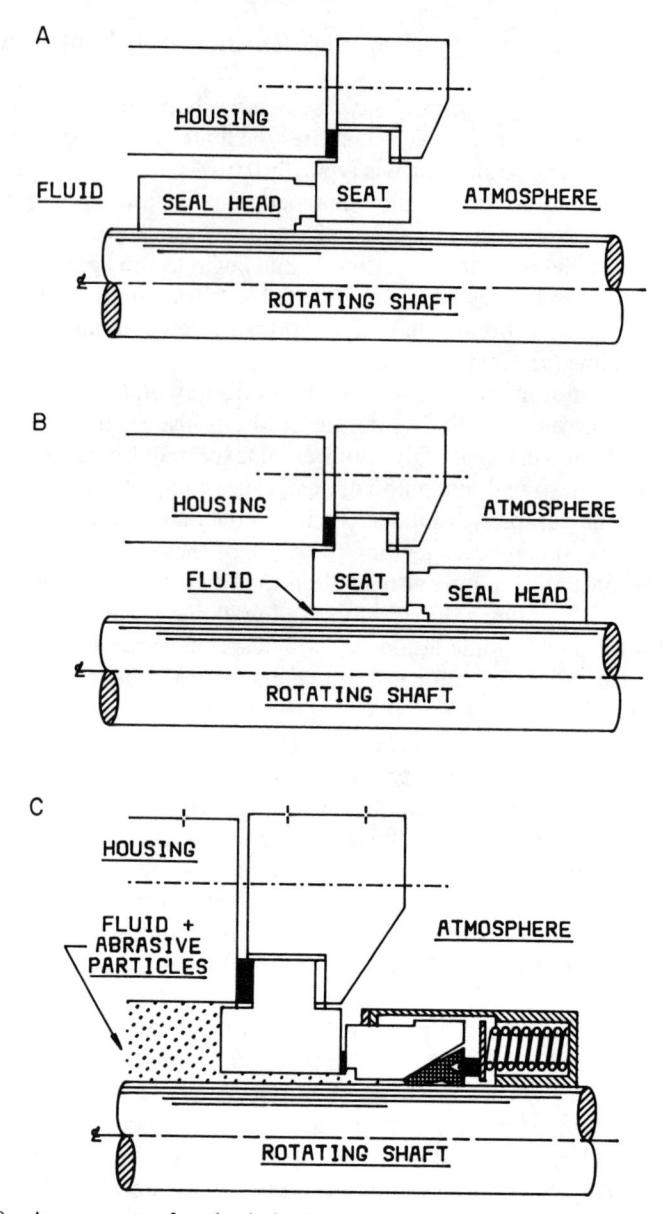

FIG. 10. Arrangements of mechanical seals. A, Single seal. It is the most common and handles most applications. B, Inside mounted seal. It operates better because it has positive lubrication; the entire seal is surrounded by fluid. C, Outside mounted seal. It is more easily accessible for maintenance and less of the seal is exposed to corrosive fluid.

Properly installed and maintained seal-less pumps, especially magnetic drive pumps, offer an economical and safe way to minimize leaks of hazardous liquids.

3. *Types of Seals*

   a. *Single Mechanical Seals.* Single mechanical seals (Fig. 10) are the most common type of mechanical seals. They provide no opportunity for a second line of defense in case the seal fails and are not recommended for hazardous or flammable service. Double mechanical seals or tandem mechanical seals should be considered for hazardous and flammable service. The failure mode of these seals is such that catastrophic releases are probably unlikely. Double and tandem mechanical seals provide a means for failure to be positively detected before it becomes serious, and the fluid being handled can be nearly totally contained. Using stuffing boxes eliminate is normally not recommended for hazardous or flammable fluids, but is often used on water or high-viscosity fluids. Packing may be necessary where the fluid is very hot or corrosive, and mechanical seals will not perform well.

   b. *Double-Seal Pumps.* With double seal pumps, as shown in Figs. 11 and 12, oil used to pressurize the space between the seals is kept at a slightly higher pressure than the process fluid being pumped (usually about 15 psi higher). There is a tiny amount of oil leakage into the process fluid under normal conditions. Oil lubricates both seals. The springs on the seals also run in oil. Running the pump without process fluid will not damage the pump as long as there is oil in the reservoir and the reservoir is under pressure. If there is seal failure, pressure on the reservoir holding the oil will force oil into the process fluid or out of the seal on the low pressure side, and the oil reservoir will show an abnormally low level. The system can be programmed to give an alarm and shut down the process and valves leading to the pump, if desired. The opportunity for major loss of the fluid being pumped is small. Whatever leakage occurs would be oil into the process fluid.

   c. *Tandem-Seal Pumps.* If no contamination of the fluid being pumped can be tolerated, a tandem seal can be used. A tandem seal is the same as two single mechanical seals in series (see Fig. 12). The seal on the process fluid side is lubricated by the fluid being pumped. There is a very small amount of process fluid leakage into the reservoir. The reservoir tank is maintained at atmospheric pressure or a little higher. A failure of the mechanical seal on the process fluid side will cause the reservoir to fill up abnormally fast. The system can be programmed to give an alarm and automatically shut down the pump and valves leading to the pump, if desired, in case of an abnormal change in the

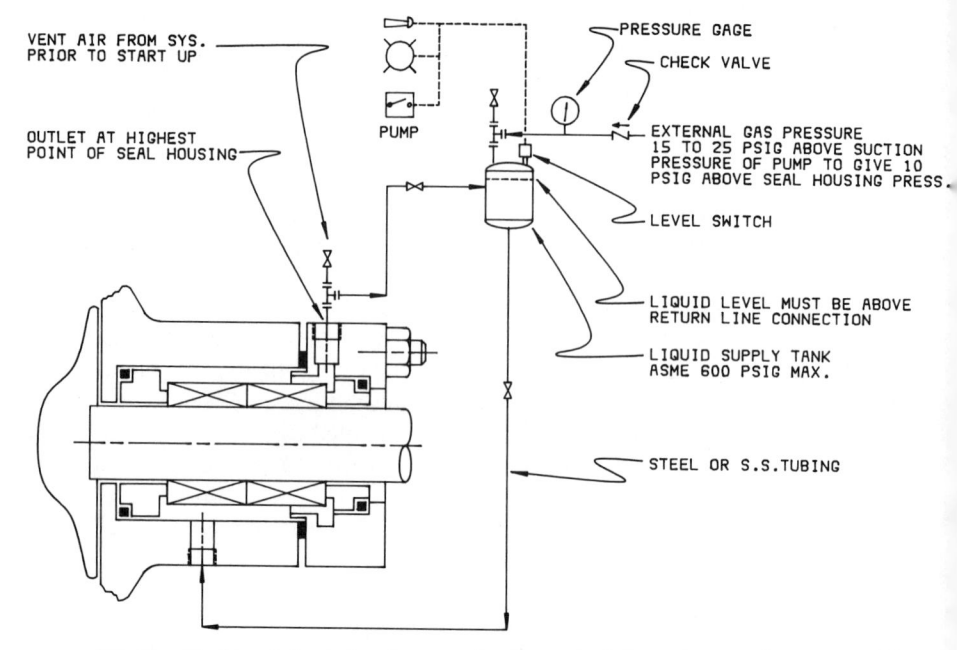

FIG. 11. Double mechanical seal accessories: Pressurized thermal convection system with liquid level switch and 1-gal ASME vessel.

reservoir level. The pump can run on the secondary seal (the normally low-pressure seal) until the system can be shut down.

A disadvantage of tandem seals is that they cannot safely be operated without process fluid because the seal on the process fluid side may run dry. Also, the springs and other internal seal mechanisms on the fluid side run in the process fluid, which may not be good in highly corrosive fluid. The big advantage of tandem seals is that they allow process fluid to be pumped without contamination from lubricating oil, and the chances for major loss of process fluid are small.

4. *Pumps in General*

There are many parts in pumps and pump seals. It is important to check all the details of internal gaskets and the internal construction of a pump to be certain these parts are compatible with the process fluids being pumped.

5. *DeadHeaded Pumps*

A deadheaded pump is a pump that is running, while full of liquid, with the inlet and outlet valves closed. Pumps should be protected against the

DOUBLE SEAL

TANDEM SEAL

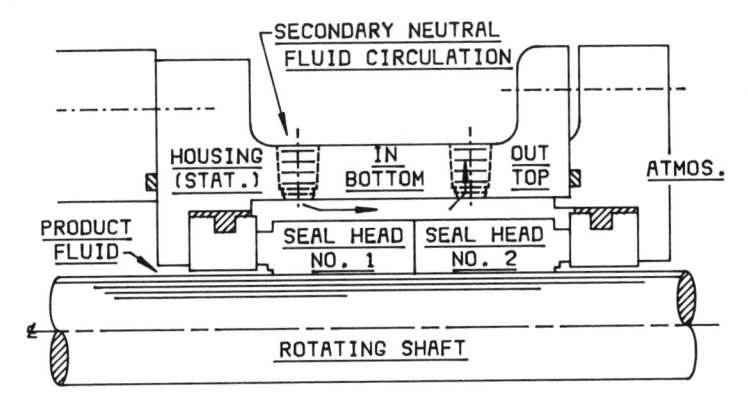

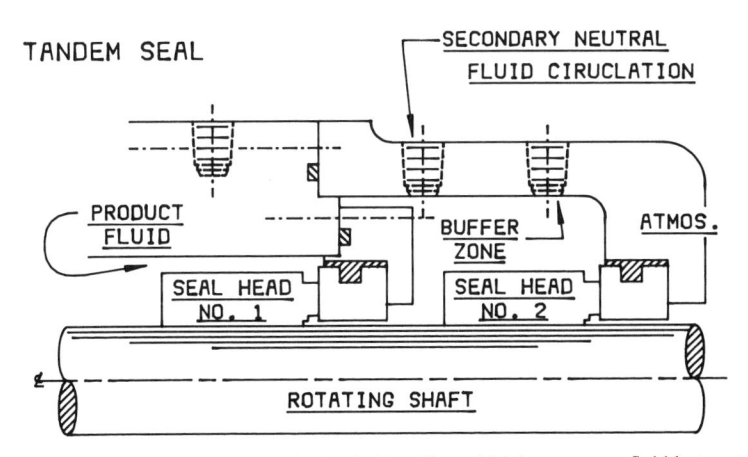

FIG. 12. Arrangements of mechanical seals. Double seal; higher pressure fluid between seals creates an "artificial environment" for seal operation. Tandem seal; two single seals with a buffer zone in between them. One backs up the other.

possibility of running deadheaded by installing a temperature device on the pump casing that can detect overheating, give an alarm, and shut the pump off. Deadheading of a centrifugal pump can result in extremely high pressures and explosions because the energy put into the liquid in the pump by the action of the impeller can cause rapid heating of the pump contents. This can result in hydraulic overpressure, high pressures caused by boiling liquid in the pump, or high pressures caused by possible chemical reactions in the pump. Water in deadheaded pumps has been the cause of several pump explosions at Dow. Other chemicals have also caused the same experience, sometimes made worse by reactions occurring within the deadheaded pump.

It has been found that in many cases, measuring amperage on the pump motor is not a very satisfactory way to detect deadheading because of the nature of pump curves. Also, the power factor of a motor operating at low capacity is generally low, which tends to obscure the actual power required if power use is deduced from amperage. For this reason, temperature devices on the pump are usually preferred as a reliable method for detecting deadheaded pumps. A positive way to measure flow from the pump may be a satisfactory method to detect and prevent dead-heading.

VI. Conclusion

This chapter addresses many of the factors to be dealt with by practicing process engineers who design chemical plants. There are many techniques that can be used to assure that a plant will be inherently safer than was ever possible before. Some of the latest knowledge available in the area of safe plant design is described. The people involved include those in management, process design, research, and manufacturing. Safety and loss prevention specialists and many other types of specialists are also of vital importance. A number of references are included for those who wish to pursue more fully any of the topics discussed.

References

Bartknecht, W., "Explosion Course Prevention, Protection," pp. 39, 55. Springer-Verlag, Berlin and New York, 1981.
Brasie, W. C., "Loss Prevention," Vol. 10, p. 135. Inst. Chem. Eng., New York, 1976.
Brasie, W. C., personal communication (1986).
Brooks, K., Rhein, R., and Bluestone, M., *Chem. Week* Sept. 24, p. 38 (1986).
"CEFIC Views on the Quantitative Assessment of Risks From Installations in the Chemical Industry." Eur. Counc. Chem. Manuf. Fed., Brussels, 1986.
"Corporate Safety and Loss Prevention." Dow Chem. Co., Midland, Michigan, 1981.
Cox, F. A., and Shiah, J. T., Gen. Pat. 1,964,213 to Goodyear Tire and Rubber Co., 1970 [C.A. Vol. 73, No. 13, 77855 (1970)].
Cromie, J., personal communication (1986).
DeHaven, E. S., *Plant Oper. Prog.* **3,** No. 1, p. 21 (1983).
Englund, S. M., *J. Chem. Educ.* No. 9, p. 59 (1982).
"EPA Asks Rules to Prevent Leaks in Underground Tanks," *New York Times* Apr. 3 (1987).
Fauske, H. K., and Leung, J., *Chem. Eng. Prog.* Aug., p. 39 (1985).
Fisher, H., *Chem. Eng. Prog.* Aug., p. 33 (1985).
Frey, K. R., personal communication (1987).

Geyer, G., personal communication (1986).
Grinwis, D., Larsen, P., and Shrock, L., personal communication (1986).
Grolmes, M. A., and Lung, J. C. (Fauske and Assoc.), *Chem. Eng. Prog.* Aug., p. 47 (1985).
Howard, W. B., and Karabinis, A. H., "Tests of Explosion Venting Buildings." Monsanto Co., St. Louis, MO, 1981.
Huff, J. E., *Loss Prev.* 7, 43 (1973).
Huff, J. E., *Repr. Symp. Loss Prev. Saf. Promotion Process Ind.*, Heidelberg, Vol. IV, p. 223. *DECHEMA*, Frankfurt am Main.
Huff, J. E., *Plant Oper. Prog.* 1, No. 4, p. 211 (1982).
Huff, J. E., *Plant Oper. Prog.* 3, No. 1, p. 50 (1984).
Huff, J. E., *Inst. Chem. Eng. Symp. Ser.* No. 85, p. 109 (1984b).
Huff, J. E., *Int. Symp. Preventing Major Chem. Accid.*, Washington, D.C., pp. 4.43–4.68, sponsored by Cent. Chem. Process Saf. Am. Inst. Chem. Eng., U.S. Environ. Agency and World Bank (1987).
Jackson, B. L., personal communication (1986).
Klein, H. H., *Plant Oper. Prog.* 1, No. 4, pp. 1–10 (1986).
Kletz, T., "Cheaper, Safer Plants or Wealth and Safety at Work," p. 49. Inst. Chem. Eng., Rugby, England, 1985a.
Kletz, T. *Design Oppor.* Sept., p. 172 (1985b).
Lees, F., "Loss Prevention in the Process Industries," 2 Vols. Butterworth, London, 1980.
Levenspiel, O., "Chemical Reactor Omnibook," pp. 3.2, 5.2. Oregon State Univ. Bookstores, Corvallis, 1979.
Liening, G., personal communication (1986a).
Liening, G., *Chem. Process.* Sept. (1986b).
"Loss Prevention Principles." Dow Chem. Co., Midland, Michigan, 1986.
Mutsakis, M. (Koch Eng. Co.), Streiff, F. A., and Schneider, G. (Sulzer Bros., Ltd., Winterhur, Switz.) *Chem. Eng. Prog.* July, pp. 42–48 (1986).
NFPA (National Fire Protection Association) 69, "Explosion Prevention Systems." NFPA, Batterymarch Park, Quincy, Massachusetts, 1986.
NFPA (National Fire Protection Association) 68, "Explosion Protection Systems." NFPA, Batterymarch Park, Quincy, Massachusetts, 1988.
Pase, G. K., *Chem. Eng. Prog.* Sept., p. 53 (1986).
Perry's Chemical Engineer's Handbook," 6th Ed., pp. 23–48. McGraw Hill, New York, 1984.
Reynolds, J. A., *Chem. Process.* 71–75 (Nov. 1989).
"Safety Standards." Dow Chem. Co., Midland, Michigan, 1982.
Sjoren, S., and Grueiro, W. (Alfa-Laval, Inc.), *Hydrocarbon Process.* Sept., p. 133 (1983).
Townsend, D. I., and Tou, J. C., *Thermochim. Acta* 37, 30 (1980).
U.S. Pat. 3,563,964 to Dow Chem. Co., 1971 [C.A. 74, No. 127,159 (1971)].
U.S. Pat. 4,024,329, Lauer, R., Rankle, R., and Dehnke, T., to Dow Chem. Co., 1977 [C.A. 87, No. 23985 (1977)].
Wensley, J. H., (Oct. 13, 1986). "Improved Alarm Management Through Use of Fault Tolerant Digital Systems," *Instrum. Soc. Am., Int. Conf. Exhibit*, Houston, Tex.

INTERACTIONS BETWEEN COLLOIDAL PARTICLES AND SOLUBLE POLYMERS

**H. J. Ploehn and
W. B. Russel**

Department of Chemical Engineering
Princeton University
Princeton, New Jersey 08544

I. Introduction

Interactions between soluble polymer and colloidal particles control the behavior of a large number of chemical products and processes and, hence, their technological viability. These dispersions have also attracted considerable scientific interest because of their complex thermodynamic and dynamical behavior—stimulated by the synthesis of novel polymers, improved optical and scattering techniques for characterization, and a predictive capability emerging from sophisticated statistical mechanical theories. Thus, the area is active both industrially and academically as evidenced by the patent literature and the frequency of technical conferences.

Polymers in colloidal systems function in a variety of ways depending on their molecular structure and concentration, the nature of the solvent, and the characteristics of the particles. Three primary types of situations exist for *homopolymers*:

(1) *Adsorption.* When each segment can adsorb reversibly onto a particle with energy comparable to kT, the chain assumes a configuration (Fig. 1b) quite different from the random coil in bulk solution (Fig. 1a). At equilibrium, bound and free segments rapidly interchange within the adsorbed chains, but bound and free chains exchange much more slowly.

(2) *Depletion.* Without adsorption the chain receives no compensation for the configurational degrees of freedom lost in approaching the surface. Hence, the center of mass generally remains a coil radius away from the surface, and the segment density at the surface falls below that in the bulk solution (Fig. 1c).

(a) random coil

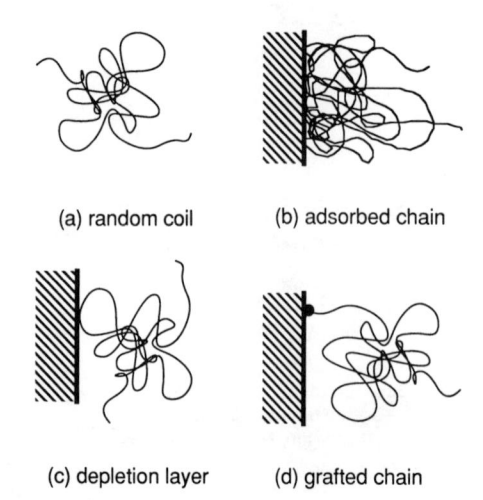

(b) adsorbed chain

(c) depletion layer

(d) grafted chain

FIG. 1. Conformation of homopolymer chains in solution: (a) Bulk solution; (b) adsorbing; (c) nonadsorbing; (d) terminally anchored.

(3) *Grafting.* A chain which does not adsorb can be tethered to the surface by one end, e.g., through a chemical reaction. The rest of the molecule then extends toward the bulk solution (Fig. 1d).

The widespread development of copolymers provides several interesting combinations of these modes of interaction:

(1) *Random copolymers* composed of strongly adsorbing and nonadsorbing groups assume different configurations than adsorbing homopolymers, e.g., with the size of loops determined by the distance between adsorbing groups rather than the balance between configurational and enthalpic contributions to the free energy (Fig. 2a). Most modern polymeric flocculants have such structures (Rose and St. John, 1985).

(2) *Diblock copolymers* consisting of soluble and insoluble parts (Fig. 2b) act much as grafted chains once they are adsorbed on the surface. However, the thermodynamics of the initial solution, consisting primarily of micelles, and the conformation of the insoluble blocks on the surface affect the coverage in ways not well understood (e.g., Munch and Gast, 1988; Marques et al., 1988; Gast, 1989). Many dispersants or polymeric surfactants are synthesized in this way (Reiss et al., 1987).

(3) Grafting hydrophobic groups onto the ends of water soluble chains forms *triblock copolymers*, which associate to form a gel in solution and adsorb reversibly onto hydrophobic portions of particle surfaces (Fig. 2c). Molecules of this sort, carefully tailored to balance the hydrophobic and

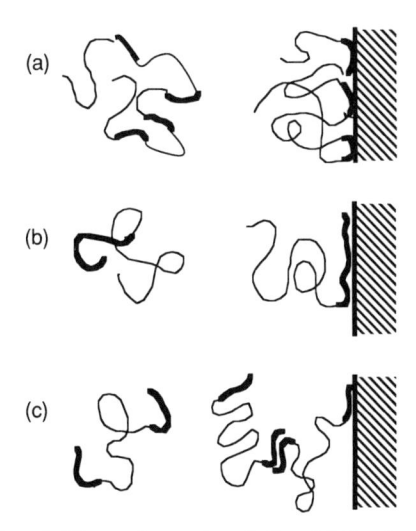

FIG. 2. Conformation of AB copolymers in solution: (a) Random copolymer; (b) diblock copolymer; (c) triblock copolymer.

hydrophilic functions, comprise the associative thickeners hailed as a revolutionary advance in the coatings industry (Sperry et al., 1987).

Even these simple classifications of the interactions between polymer and colloidal particles present a wide variety of possibilities, suggesting the importance of understanding the effect of the macromolecular and particle characteristics on specific macroscopic processes. Unfortunately, the linkage is incomplete in most cases; however, considerable qualitative understanding has evolved relating the interparticle potential to the behavior of colloidal systems. In addition, direct measurements of forces between surfaces exposed to or treated with well-characterized polymers have begun to define the phenomena and provide quantitative data for testing theories. Thus, we address the more limited issue of quantifying this potential, as affected by the various modes of interaction. Then in the final section of this review, we discuss the effects on both the quiescent state of colloidal dispersions, i.e., stabilization, flocculation, phase separation, and dynamics such as the rheology.

This review is intended to complement those of Cohen Stuart et al. (1986) and de Gennes (1987). The former details experimental techniques available for probing polymer–particle interactions and the lattice, i.e., mean field, theories that predict, via numerical solutions, segment-density profiles and interaction potentials. The latter constructs a simple and elegant picture of the same phenomena through scaling theories developed for semidilute solutions.

Here we steer a middle course, emphasizing analytical results extracted from the mean-field theories, in context with the lattice and scaling approaches, and explaining the relationship to macroscopic phenomena. Our treatment draws on Russel et al. (1989) in some places.

II. Polymer Solution Thermodynamics

A. General Features

The preceding section illustrates the variety of phenomena that may be observed in polymer–colloid–solvent mixtures. Polymer dissolved in a colloidal suspension is in some ways similar to ionic solutes responsible for electrostatic effects. Interactions between colloidal particles and polymer generate nonuniform distributions of polymer throughout the solution. Particle–particle interactions alter the equilibrium polymer distribution, producing a force in which sign and magnitude depend on the nature of the particle–polymer interaction. The major difference between polymeric and ionic solutions lies in the internal degrees of freedom of the polymer. Thus, a complete treatment of particle–polymer interactions requires detailed consideration of the thermodynamics of polymer solutions.

A large body of theoretical (Flory, 1953; Yamakawa, 1971; Lifshitz et al., 1978; de Gennes, 1979; Freed, 1987) and experimental (Brandrup and Immergut, 1975) results demonstrates that many of the properties of polymers in solution display universal behavior as a consequence of the molecules' chainlike architecture. Measurements on a length scale comparable to the size of the polymer chain are insensitive to the details of atomic arrangements or interactions. Polymer properties that depend primarily on gross molecular structure include the radius of gyration (R_g) and the second virial coefficient (A_2) in the series expansion of the solution osmotic pressure in powers of polymer concentration. Under many conditions R_g and A_2 vary as KM^a where M is the polymer molecular weight. In general, K and a differ for R_g and A_2, and depend on the characteristics of the materials. Nonetheless, universal theories successfully predict many solution properties over a wide range of conditions, suggesting the functional relationships (such as power-law) scaling) among key dimensionless groups formed from the parameters characterizing the solution.

Universal models of polymer solutions attempt to describe a variety of large-scale properties with a minimum number of phenomenological parameters. Some theories (Flory, 1969) predict these parameters through microscopic models of bond geometry and interactions, and difficult but unavoidable procedure when the characteristic length scale in the solution is comparable to the bond length. In bulk polymer, for example, molecular interactions influence many macroscopic properties (Bovey, 1982) and necessitate detailed models. Fortunately, most of the features relevant to interactions between colloidal particles and soluble polymer can be predicted by theories that fix a few phenomenological parameters through independent measurements. These polymer solution theories only require the polymer concentration, the contour length of the molecule, a measure of its flexibility, and a minimal characterization of interactions among different parts of the molecule.

Universality allows the detailed, monomer-level description of the polymer to be replaced by a simpler, equivalent model. The most common random-walk model idealizes the polymer molecule as a freely jointed chain of n segments of length l. If the segments were aligned, the chain would be fully extended with a mean-square end-to-end distance $\langle r^2 \rangle = (nl)^2$. Thermodynamically, this state is highly improbable. The most likely configuration when segments do not interact is a random walk with $\langle r^2 \rangle_0 = nl^2$ where the subscript 0 denotes ideality. The real molecule and this idealized chain must have the same contour length and mean-square end-to-end distance. With l_b and m_b as the average bond length and molecular weight per bond, this leads to

$$\frac{M}{m_b} l_b = nl \qquad (1)$$

and

$$\langle r^2 \rangle_0 \equiv \frac{C_\infty M}{m_b} l_b^2 = nl^2, \qquad (2)$$

where C_∞ is the characteristic ratio defined by the first equality in Eq. (2). Values of l_b, m_b, and C_∞ have been measured for many systems (Brandrup and Immergut, 1975); some examples are given in Table I. The segment length l measures the chain flexibility in such one-parameter models of ideal polymer solutions.

In real molecules, segments on distant portions of the chain do interact because of their physical volume and short range attractions such as van der Waals forces. The excluded volume parameter

$$v \equiv \int [1 - e^{-V(r)/kT}] dr, \qquad (3)$$

derived from the mean (solvent-mediated) interaction potential $V(r)$ between a pair of segments separated by r, represents the effective volume that one

TABLE I

PROPERTIES OF POLYMER IN SOLUTION[a]

	l_{ϕ_b} (nm)	m_{ϕ_b} (10^{-3} kg/mol)	$\tilde{v}$ (10^{-3} m^3/kg)	C_∞	l (nm)	m (10^{-3} kg/mol)
Poly(oxyethylene)	0.148	14.7	0.79	4.1	0.60	60
Poly(12-hydroxy-stearic acid)	0.150	20	—	6.1	0.91	122
Poly(dimethyl siloxane)	0.162	37	0.98	5.2	0.84	192
Polystyrene	0.154	52	0.95	9.5	1.46	494

[a] Taken from Brandrup and Immergut (1975).

segment excludes to all others. The ideal or "theta" state occurs at a temperature $T = \theta$ at which $v = 0$. Expanding Eq. (3) around $T = \theta$ gives

$$v \approx Kl^3\left(1 - \frac{\theta}{T}\right), \tag{4}$$

where the constant K may be positive or negative depending on the variation of $V(r)$ with T. For $K > 0$, (4) indicates that $T > \theta$ in good solvents ($v > 0$), while $T < \theta$ in poor solvents ($v < 0$). Theoretical treatments of an isolated chain (e.g., Yamakawa, 1971) indicate that the chain approximates an ideal random walk when $-1 < vn^{1/2}/l^3 < 1$, but that excluded volume "swells" the chain such that $\langle r^2 \rangle \approx n^{6/5}l^{4/5}v^{2/5}$ when $vn^{1/2}/l^3 > 1$. In poor solvents (Moore, 1977a; Lifshitz et al., 1978), chains contract so that $\langle r^2 \rangle \approx n^{2/3}l^2$, but a complete treatment of this coil–globule transition requires consideration of more than pair interactions. The two phenomenological parameters, l and v, provide an adequate description of the behavior of dilute solutions for $v > 0$ and appear in the dimensionless universal groups $\langle r^2 \rangle/nl^2$ and $vn^{1/2}/l^3 = vn^2/\langle r^2 \rangle_0^{3/2}$, such that $\langle r^2 \rangle/nl^2 = f(vn^{1/2}/l^3)$, in most two-parameter theories (Yamakawa, 1971).

As the segment density ρ increases from the dilute limit (or as coils collapse for $v < 0$), higher-order interactions (three-body, four-body, ...) become important. The fraction of space excluded by chains may be expressed by the expansion $\rho v/2 + \rho^2 w/3 + \cdots$, where $v, w, \ldots$ are the corresponding many-body cluster integrals. As ρ increases, an accurate representation of solution properties requires either more phenomenological parameters or closure approximations, which express the higher-order interactions in terms of v and w. The simplest closure assumes that segment interactions can be represented by a mean field averaged over some region of space. Such mean-field theories are most accurate for concentrated solutions in which higher-order interactions eliminate pair correlations.

Beyond the *dilute* limit lies a range of concentrations in which the solution properties are adequately characterized by n, l, v, and w. The *semidilute* regime begins when the average polymer concentration becomes comparable to that within an isolated coil, i.e., when chains begin to overlap and intermolecular interactions become important. As concentration increases further, the intermolecular interactions dominate the intramolecular interactions, marking the transition from semidilute to *concentrated* systems. Within the semidilute regime, Daoud and Jannink (1976) and Schaefer (1984) map several distinct domains of qualitatively different behavior on a temperature–concentration (T–c) diagram.

A schematic T–c diagram is shown in Fig. 3 for fixed values of n, l, and w. Note that the excluded volume v is related to T through Eq. (4), and the polymer volume fraction φ is related to mass concentration c and chain number density ρ/n through

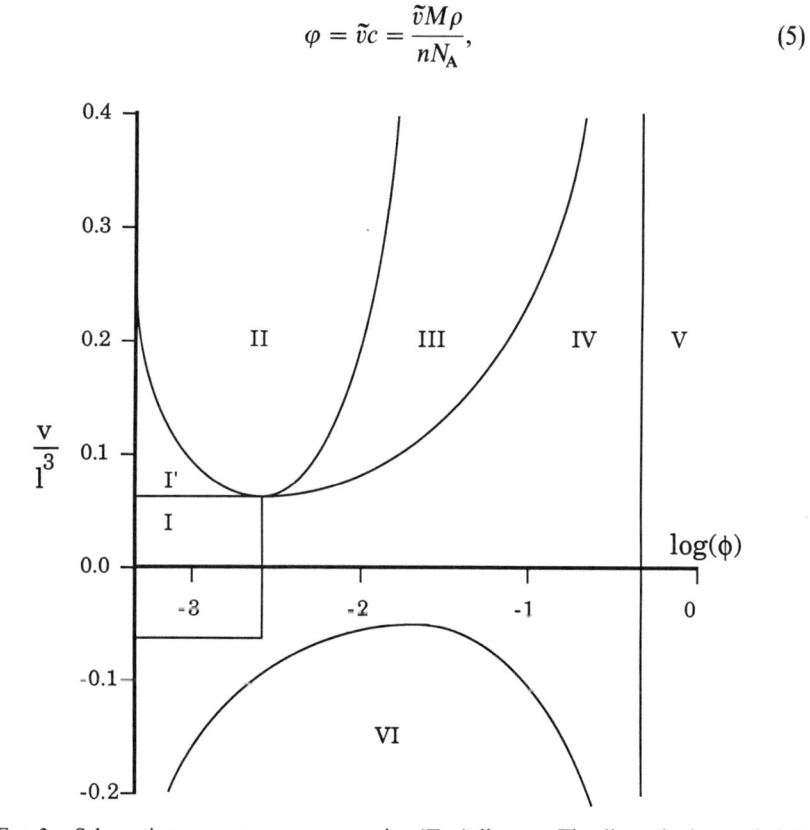

$$\varphi = \tilde{v}c = \frac{\tilde{v}M\rho}{nN_A}, \tag{5}$$

FIG. 3. Schematic temperature–concentration (T–c) diagram. The dimensionless excluded volume v/l^3 is related to T by Eq. (4); φ is proportional to c via Eq. (5). The various regimes are described in the text.

where $\tilde{v}$ is the polymer specific volume, and N_A is Avogadro's number. In region I, chains assume isolated, ideal random walks ($\langle r^2 \rangle_0 \approx nl^2$). Intrachain-excluded volume swells isolated chains ($\langle r^2 \rangle \approx n^{6/5} l^{4/5} v^{2/5}$) in region I'. Region V denotes the concentrated regime where higher-order interactions ($\geq$ three-body) become important. On the poor-solvent side of the diagram, phase separation of the polymer—solvent mixture occurs in region VI. The most significant regions for polymer—colloid interactions, labeled II, III and IV, have been designated (Schaefer, 1984) as the semidilute-good, -marginal, and -theta regimes, respectively.

Within the semidilute-good regime, chains swell because of intrachain excluded volume, but also they overlap and interact. The resulting entangled network of chains has a correlation length ξ as introduced by Edwards (1966). On length scales greater than ξ, random interchain interactions screen intrachain-excluded volume, and chains are ideal, but excluded volume still causes the osmotic pressure to increase nonlinearly with φ. As φ increases, ξ decreases and chains relax toward the ideal state. In the semidilute-marginal regime, excluded volume only weakly perturbs chain configurations so that chains are essentially ideal on all length scales. As φ increases still further, or as v decreases, three-body interactions dominate two-body terms, producing the semidilute-theta regime.

The crossover curves between each pair of regions, detailed by Schaefer (1984), should be interpreted as approximate delimiters rather than sharp boundaries. Within each region, different physical phenomena control the solution's thermodynamic behavior, and so different theoretical approaches have developed. Scaling theory (de Gennes, 1979) predicts trends which agree with experimental data for semidilute-good solutions. Renormalization-group theory (Freed, 1987) supplements scaling theory by providing numerical coefficients and extending predictions beyond the semidilute-good regime within a universal framework. For semidilute-marginal and -theta solutions, mean-field theories have a long history (Yamakawa, 1971) and take a variety of forms. Of these, the self-consistent field (SCF) method has proven most useful for modelling polymer solutions as well as polymer interfaces.

B. Scaling Theory

The concepts of scaling theory are based on the analogy between polymer chain statistics and the properties of fluids and magnets near critical conditions (de Gennes, 1972, 1979, 1987; Des Cloizeaux, 1975). Although most expositions introduce critical phenomena in the context of magnetic materials, the phenomenology of liquid–gas phase transitions is equally valid. In any homogeneous phase, the temperature T and pressure P determine the fluid's properties, such as the density ρ, compressibility $(\partial \rho/\partial P)_T$, and heat capacities C_P and C_V, through a suitable equation-of-state. On the boundary between two phases, however, the fluid properties are not uniquely determined. The liquid–gas phase boundary, represented by a T, P curve on a phase diagram, ends at the critical point (T_c, P_c), where the densities of the two phases become equal to ρ_c. Near the critical point, large fluctuations of the local density $\rho(\mathbf{r})$ from the mean value ρ are observed. Correlations of the fluctuations, measured by neutron scattering through the Fourier transform,

$$G(q) = \int \langle [\rho(\mathbf{r}) - \rho][\rho(\mathbf{0}) - \rho] \rangle e^{-i|\mathbf{q}\cdot\mathbf{r}|}\, d\mathbf{r},$$

with $\langle \cdots \rangle$ denoting an average over all possible configurations (states) of the system, are found to be peaked at $q = 0$ with the width of the peak defined through

$$\xi^{-2} \equiv -\frac{1}{2G(0)}\left[\frac{d^2 G}{dq^2}\right]_{q=0}.$$

Physically, ξ represents the correlation length of the density fluctuations near the critical point. Many measurements demonstrate that for T near T_c,

$$\xi \approx |T - T_c|^{-\nu}, \tag{6}$$

where ν is a critical exponent. Measurements for a variety of systems (liquid–gas phases, liquid mixtures, alloys, magnetic materials) invariably indicate $\nu \approx \frac{3}{5}$, demonstrating the universality of ν.

For polymer solutions, de Gennes (1972) first identified $|T - T_c|$ with $1/n$ and with the root–mean–square end-to-end distance for an isolated chain. Thus, an expression analogous to Eq. (6) is

$$\langle r^2 \rangle^{1/2} \approx n^\nu l, \tag{7}$$

valid for dilute solutions in the limit $n \to \infty$. For ideal chains, $\nu = \frac{1}{2}$ results from the classic analyses of random walks (e.g., Chandrasekhar, 1943). The value $\nu \approx \frac{3}{5}$ for good solvents (0.588 is the most accurate value indicated) has been verified theoretically (Flory, 1953; Edwards, 1965; Yamakawa, 1971; Freed, 1987), numerically (Domb, 1963, 1969), and experimentally (Flory, 1969).

The concentration dependence follows through heuristic arguments (de Gennes, 1979) or formal mathematical definitions (Des Cloizeaux, 1975; Kosmas and Freed, 1978). The crossover from dilute to semidilute behavior in good solvents occurs at a segment density $\rho^* \approx n/\langle r^2 \rangle^{3/2} \approx n^{-4/5} v^{-3/5} l^{-6/5}$. For $\rho > \rho^*$, the solution resembles an entangled network with a correlation length $\xi(\rho)$, which is independent of n. At the crossover, the correlation length must vary continuously so that $\xi = \langle r^2 \rangle^{1/2}$ at $\rho = \rho^*$. In terms of $\varphi \equiv \rho l^3$ [not the φ defined in Eq. (5)], scaling theory derives $\xi(\varphi)$ by forming the dimensionless groups $\xi/\langle r^2 \rangle^{1/2}$ and $\rho/\rho^* = \varphi/\varphi^*$, and postulating the functional relation

$$\frac{\xi(\varphi)}{\langle r^2 \rangle^{1/2}} \approx f(\varphi/\varphi^*) \approx \left[\frac{\varphi}{\varphi^*}\right]^m \tag{8}$$

for φ/φ^* large (Freed, 1987). Using

$$\langle r^2 \rangle^{1/2} \approx n^{3/5} v^{1/5} l^{2/5}$$

and

$$\varphi^* \approx n^{-4/5} v^{-3/5} l^{9/5},$$

the value $m = -\tfrac{3}{4}$ gives ξ independent of n as

$$\xi(\varphi) \approx \varphi^{-3/4} v^{-1/4} l^{7/4} \tag{9}$$

for $\varphi > \varphi^*$. Data by Daoud et al. (1975) and Wiltzius et al. (1983) shown in Figs. 4 and 5 confirm this behavior. Further comparisons of scaling-theory predictions and neutron-scattering measurements for moderate ranges of φ and T are presented by Cotton et al. (1976) in terms of a $T - c$ diagram as proposed by Daoud and Jannink (1976).

The osmotic pressure Π is the thermodynamic function most accessible to experimental measurement. Dilute solutions of chains exhibit ideal behavior

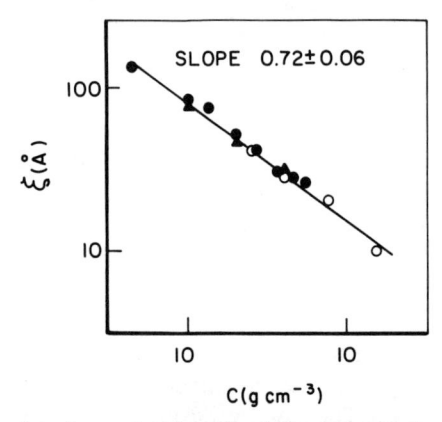

FIG. 4. Correlation length as a function of polymer concentration measured by neutron scattering by Daoud et al. (1975). Closed symbols denote polystyrenes of two different molecular weights in deuterated benzene; open symbols denote deuterated polystyrenes of two molecular weights in carbon disulfide.

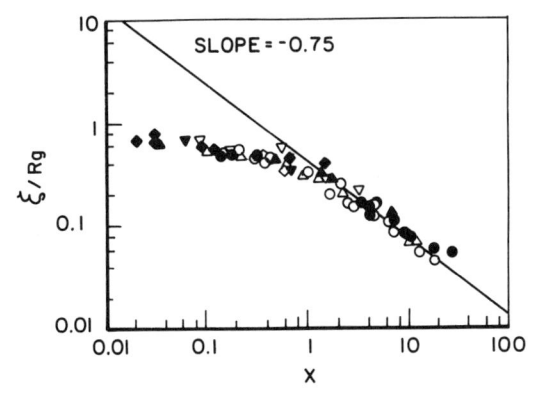

FIG. 5. Correlation length, scaled on R_g, as a function of reduced concentration $X \approx c/c^*$ as measured through static light-scattering by Wiltzius et al. (1983). Symbols denote various molecular weight polystyrenes in toluene and methyl ethyl ketone at 25°C.

with $\Pi = \rho kT/n$. At semidilute concentrations, Π has the general form

$$\frac{\Pi}{kT} = \frac{\rho}{n} f\left[\frac{\rho}{\rho^*}, \frac{v}{l^3}, \frac{l_b}{l}\right] = \frac{\varphi}{nl^3} f\left[\frac{\varphi}{\varphi^*}, \frac{v}{l^3}, \frac{l_b}{l}\right], \quad (10)$$

with f an unknown function and $\varphi \equiv \rho l^3$. The scaling hypotheses summarized by Freed (1987) require that f depend only on large-scale variables so that

$$f\left[\frac{\varphi}{\varphi^*}, \frac{v}{l^3}, \frac{l_b}{l}\right] = f_1\left[\frac{\varphi}{\varphi^*}\right] \approx \left[\frac{\varphi}{\varphi^*}\right]^q,$$

with the power-law scaling again assumed for φ/φ^* large. With $\varphi^* \approx n^{-4/5} v^{-3/5} l^{9/5}$, Eq. (10) becomes

$$\frac{\Pi}{kT} \approx \frac{\varphi}{nl^3}\left[\frac{\varphi}{\varphi^*}\right]^q \approx \varphi^{q+1} n^{4q/5-1} l^{-3}. \quad (11)$$

Since Π must be independent of n, $q = \frac{5}{4}$ and

$$\frac{\Pi}{kT} \approx \varphi^{9/4} v^{3/4} l^{-21/4} \approx \frac{1}{\xi^3}. \quad (12)$$

Thus, Π scales as the thermal energy per correlation volume. Neutron scattering (Daoud et al., 1975) and laser-light scattering (Wiltzius et al., 1983) yield $\partial \Pi/\partial \varphi$ in the limit of zero scattering vector support and the scaling $\partial \Pi/\partial \varphi \; \varphi^{5/4}$ for $\varphi > \varphi^*$. Direct measurements of Π (Noda et al., 1981), reproduced in Fig. 6, agree with the predicted scaling of Eq. (12).

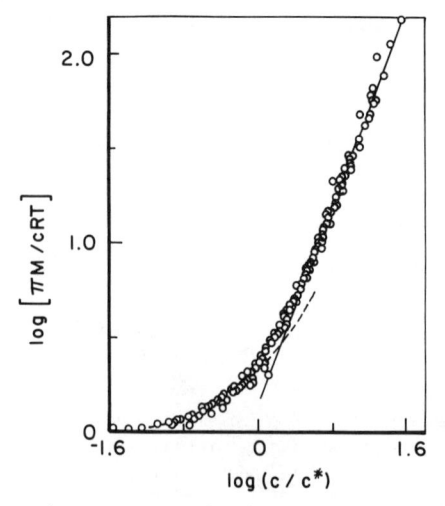

FIG. 6. Reduced osmotic pressure vs. reduced concentration, plotted on a log–log scale, measured through osmometry by Noda et al. (1981). Symbols denote various molecular weights of poly(α-methylstyrene) in toluene at 25°C. The solid line has a slope of 1.32.

C. Renormalization-Group Theories

The polymer–magnet analogy gives scaling theory its theoretical foundation but restricts its applicability in describing polymer solutions. Still, the formal mathematical similarity is compelling and prompted efforts (Edwards, 1975; Moore, 1977b; Moore and Al-Noaimi, 1978; Des Cloizeaux, 1980a,b,c, 1981; Oono and Freed, 1981) to apply renormalization group (RG) methods, developed for studies in condensed-matter physics, to the problems of polymer solutions. RG methods produce the asymptotic forms of solution properties with numerical prefactors and expressions in the crossover regions. By mathematically connecting the field theoretic models of polymer statistics with the concepts of universality and dimensional scaling, RG theory provides a systematic approach for predicting universal behavior throughout the dilute and semidilute regimes. The purpose here is to introduce some of the fundamental concepts and recent results; Freed's text (1987) is a far more complete and detailed reference.

Several varieties of RG theory are reviewed in extant texts (e.g., Stanley, 1971; Ma, 1976; Ziman, 1979; Freed, 1987). Although the details differ, all approaches share the basic operations of coarse-graining and scaling. Coarse-graining introduces phenomenological parameters that characterize but obviate fine details, thus simplifying the description and facilitating calculations. For example, sequences of real monomers become statistical segments of length l. Scaling formulates a minimal set of phenomenological parameters through dimensional analysis and relates regions of the parameter space through transformations such as rotation, translation, reflection, or more complicated symmetries. Thus, the self-similarity (translational symmetry) of semidilute solutions allows ξ to be related to l. Beginning with a set of parameters characterizing one model (e.g. n, l, v, w), a sequence of scaling and coarse-graining operations leads to a new form represented by a set of renormalized parameters. Repeating the procedure generates a series of models belonging to the same *universality class* which comprise the *renormalization group*. Renormalization ends when the model becomes invariant to transformation, i.e., when an asymptotic limit is reached. Polymer properties can, in principle, be evaluated more easily in such limits.

In semidilute solutions, chain interactions are strongly coupled, producing significant density fluctuations. One successful RG technique (Edwards, 1966, 1975; Muthukumar and Edwards, 1982) renormalizes the segment length l and the segment–segment interaction $v\delta r$ into l_1 and Δr in order to define an effective single chain in an average excluded-volume field. The new segment length is defined by

$$l_1 \equiv \frac{\langle r^2 \rangle}{nl}. \tag{13}$$

The excluded-volume field $\Delta(\mathbf{r})$

$$\Delta(r) \equiv v\left[\delta(r) - \frac{1}{4\pi\xi^2 r}\exp(-r/\xi)\right] \tag{14}$$

contains the screening length ξ equivalent to the Edwards (1966) correlation length. Properties of the effective single chain are extracted by making $\Delta(r)$ self-consistent via the "method of random fields" and then solving in the asymptotic limit of highly-stretched chains (Edwards, 1966, 1975).

Edwards and co-workers (Edwards and Jeffers, 1979; Edwards and Singh, 1979) adapted these techniques to determine chain dimensions in semidilute solutions. Subsequently, Muthukumar and Edwards (1982) formulated RG expressions for the solution free-energy and derived Π explicitly in the low and high concentration limits. The general expression for the free-energy density relative to that of an infinitely dilute solution is

$$\frac{A}{VkT} = -\frac{9}{16\pi}\frac{vn\rho}{ll_1\xi} - \frac{1}{24\pi\xi^3} + v\rho^2, \tag{15}$$

where l_1 and ξ satisfy

$$l_1^3\left[\frac{1}{l} - \frac{1}{l_1}\right] = \frac{\alpha v\xi}{l^2} \tag{16}$$

and

$$\frac{1}{\xi^2} = \frac{6v\rho}{ll_1\left[1 + \dfrac{27v\xi}{8\pi l^2 l_1^2}\right]}, \quad (17)$$

where α is a numerical coefficient. For $\rho \to 0$ and $n \to \infty$, Muthukumar and Edwards find

$$\frac{\Pi}{kT} = \frac{\rho}{n} + \frac{40\pi}{243}\left[\frac{16\pi\alpha^3}{9}\right]^{1/4} v^{3/4} l^{3/2} \rho^{9/4}, \quad (18)$$

with the $\rho^{9/4}$ dependence confirming the scaling-theory result. In the limit $\rho \to \infty$,

$$\frac{\Pi}{kT} = \frac{\rho}{n} - \frac{5\sqrt{6}}{32\pi} \frac{(v\rho)^{3/2}}{l^2} + v\rho^2, \quad (19)$$

conforming to the ρ^2 dependence expected for semidilute-marginal solutions where pair interactions dominate the solution free-energy (Schaefer, 1984). The osmotic pressure at intermediate concentrations follows from the free energy through solution of Eqs. (16) and (17) for l_1 and ξ and partial differentiation of Eq. (15). Fixing the coefficient α requires a measurement of $\langle r^2 \rangle$ at a known value of v; then

$$l_1^{5/2}\left[\frac{1}{l} - \frac{1}{l_1}\right] = \frac{\alpha v(nl)^{1/2}}{2\sqrt{6\pi l^2}}, \quad (20)$$

which, along with Eq. (13), gives α.

Another RG analysis (Des Cloizeaux, 1980a,b, 1981; Oono and Freed, 1981; Oono et al., 1981; Ohta and Oono, 1982) produces a closed-form expression for Π. With one independent measurement to fix A_2 in the dilute limit, Ohta and Oono (1982) define $X \equiv \frac{16}{9}(M/N_A)A_2\rho$ as the universal concentration variable (Wiltzius et al., 1983) and derive

$$\frac{\Pi}{\rho kT} = 1 + \frac{X}{2}\exp\left\{\frac{1}{4}\left[\frac{1}{X} + \left(1 - \frac{1}{X}\right)\ln(X)\right]\right\}, \quad (21)$$

with the scaling-theory trend ($\Pi \approx X^{9/4}$) emerging as X grows large. Comparisons with the data of Noda et al. (1981) (Fig. 6), from Ohta and Oono (1982) (Fig. 7), and from Wiltzius et al. (1983) (Fig. 8) demonstrate the validity of Eq. (21), at least for dilute and semidilute conditions. Since the Muthukumar and Edwards expression, Eq. (18), pertains to semidilute and concentrated conditions, the two are complementary.

RG theories are currently being extended into the semidilute-marginal and -theta regimes. Results (Freed, 1985; Bawendi et al., 1986) remain limited and

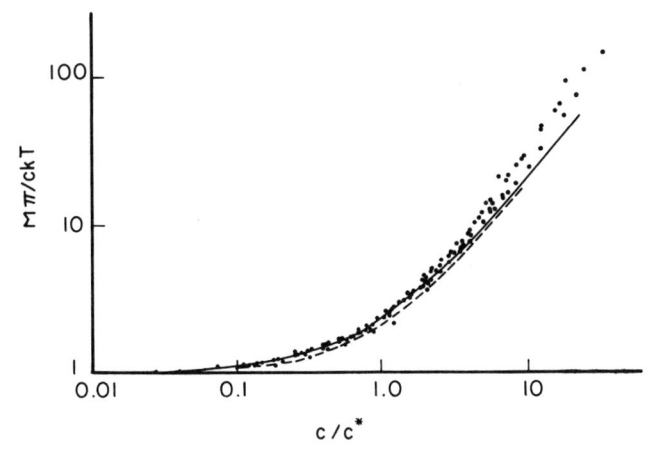

FIG. 7. $\Pi/\rho kT$ vs. reduced concentration plotted on a log–log scale. Symbols denote the data of Noda et al. (1981) as in Fig. 6. Solid curve, which has been shifted horizontally to fit the data, is the prediction of Eq. (21). Reproduced from Ohta and Oono (1982).

restricted to the vicinity of $T = \theta$ where $v \approx 0$ but $w > 0$. The complexity of calculations involving three-body interactions is daunting (Freed, 1987, Chap. 11), but may eventually lead to thermodynamic properties. At present, RG theory provides the best hope of describing polymer solutions within a unified, universal framework.

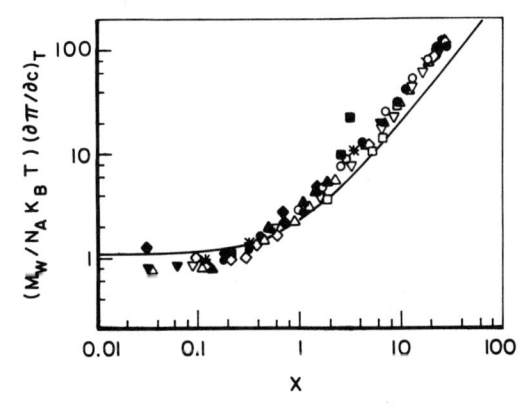

FIG. 8. Dimensionless reciprocal compressibility vs. reduced concentration plotted on a log–log scale, measured through static light-scattering by Wiltzius et al. (1983). Symbols denote data for various molecular weight polystyrenes in toluene and methyl ethyl ketone at 25°C. The solid curve is the prediction derived from Eq. (21); the proportionality of X to c is fixed by an independent measurement so the curve requires no adjustable parameters.

D. Flory Mean-Field Theories

The first mean-field theories, the lattice models, are typified by the Flory–Huggins model. Numerous reviews (see, e.g., de Gennes, 1979; Billmeyer, 1982; Forsman, 1986) describe the assumptions and predictions of the theory; extensions to polydisperse and multicomponent systems are summarized in Kurata's monograph (1982). The key results are reiterated here.

Space is discretized into a lattice of sites occupied by either segments or solvent molecules. Each site has z nearest neighbors. The number of Flory–Huggins segments per chain, $r = \tilde{v}M/V_s$, equals the ratio of polymer ($\tilde{v}M$) and solvent (V_s) molar volumes. Thermodynamic properties are computed by filling a unit volume of the lattice with a total of ρ/r chains of r segments and ρ_s solvent molecules. As a chain is placed in the lattice, the probability of inserting the ith segment is $(z-1)(1-p_i)$, since $(z-1)$ sites are available adjacent to segment $i-1$ (z neighbors minus one site occupied by segment $i-2$), and $1-p_i$ is the probability of a particular adjacent site being vacant. Flory (1942) and Huggins (1942) replace the local p_i with a mean $\langle p_i \rangle$ averaged over the lattice, creating the mean field. Thus, the theory is restricted to concentrated solutions in which density fluctuations and correlations are unimportant.

Manipulation of combinatorial statistics leads to the Flory–Huggins free energy of mixing (per unit volume)

$$\frac{\Delta f_M}{VkT} = \frac{\rho}{n}\ln(\varphi) + \rho_s \ln(1-\varphi) + \rho_s \varphi \chi \tag{22}$$

where χkT is the enthalpy of mixing a segment with an excess of solvent. The Gibbs and Helmholtz free energies of mixing both equal Δf_M since $\Delta V_M \equiv 0$. With $\varphi = \rho/(\rho + \rho_s)$ and $\Pi \equiv -[\partial(\Delta f_M V)/\partial V]_T$, a simple calculation produces

$$\frac{\Pi V_s}{N_A kT} = -[\ln(1-\varphi) + (1-1/r)\varphi + \chi\varphi^2] \tag{23}$$

$$= \frac{\varphi}{r} + (1-2\chi)\frac{\varphi^2}{2} + \frac{\varphi^3}{3} + \cdots, \tag{24}$$

allowing identification of $v/l^3 \approx (1-2\chi)$.

In practice, the Flory–Huggins theory fails to predict many features of polymers solutions, either qualitatively or quantitatively, but remains widely used because of its simplicity. The Flory parameter χ, assumed to be constant, often increases with φ. Furthermore, χ, an interaction-energy scaled on kT, often exhibits a more complicated temperature dependence than $1/T$ (Flory, 1970). Such behavior stems from energetic effects, such as directional polar interactions, which preferentially orient segments. Reinterpretation of χ as a composite of enthalpic and entropic factors (Flory, 1970) rationalizes these variations; Eq. (22) then provides a convenient means of correlating data, but the exercise does little to suggest improvements of the theory. As a mixture theory, the analysis also ignores pure component properties.

Corresponding-states theory (Prigogine, 1957; Flory, 1970) incorporates features of the pure component properties and liquid structure in the mixture equation-of-state, producing nonzero values of ΔV_m and contributing enthalpic and entropic terms beyond those in Flory–Huggins theory. The theory assumes that all pure components and mixtures obey the same universal equation-of-state, e.g., (Flory, 1970)

$$\frac{\tilde{\Pi}\tilde{V}}{\tilde{T}} = \frac{\tilde{V}^{1/3}}{\tilde{V}^{1/3} - 1} - \frac{1}{\tilde{V}\tilde{T}} \tag{25}$$

where $\tilde{\Pi}$, $\tilde{V}$, and $\tilde{T}$ are Π, V, and T reduced with parameters fixed by measurements for pure components. Further theoretical efforts, often employing lattice fluid-models (e.g., Sanchez and Lacombe, 1978), postulate mixing rules that combine the pure component parameters in mixture reduction parameters. The corresponding-states approach is generally successful in predicting mixture properties and phase behavior but requires a significant investment to determine pure component properties.

Another effort (Dickman and Hall, 1986) aims to remove the artificiality of lattice geometry by transposing the concepts of Flory–Huggins theory to continuous space. Calculations for athermal solvents ($\chi = 0$) indicate that the resulting "generalized" Flory theory predicts $\Pi/\rho kT$ in agreement with computer-simulation data for short chains of hard disks and spheres, whereas the lattice-based Flory–Huggins theory gives a severe underestimate. The Dickman–Hall theory has not yet been extended to other values of χ, however.

E. SELF-CONSISTENT MEAN-FIELD THEORIES

Like all mean-field theories, SCF theories replace the detailed, configuration-dependent interaction potentials with a mean potential averaged over the distribution of molecular configurations. Unlike other mean-field theories, SCF theory explicitly calculates the mean field by accounting for the polymer chain statistics. This field, in turn, controls the distribution of polymer configurations: Hence the term "self-consistent."

The classic SCF analyses (Edwards, 1965; de Gennes, 1969, 1979; Freed, 1972; Helfand, 1975c) define the number of configurations available to a subchain of $s \leq n$ contiguous segments which begins at $\mathbf{r}'$ and ends at $\mathbf{r}$ as

$G(\mathbf{r}, \mathbf{r}'; s)$. For a complete chain of n segments, integrating over $\mathbf{r}$ and $\mathbf{r}'$ gives the configuration integral

$$W = \iint G(\mathbf{r}, \mathbf{r}'; n) \, d\mathbf{r} \, d\mathbf{r}', \qquad (26)$$

which acts as the partition function for individual chains. The normalized function G/W is a probability density.

The configuration of a polymer chain is analogous to the path $\mathbf{r}(s)$ of a diffusing particle with the segment rank s replacing time. At equilibrium, configurations follows the Boltzmann distribution

$$G(\mathbf{r}, \mathbf{r}'; s) = \exp\left[-\int \beta U[\mathbf{r}(s')] \, ds'\right] \qquad (27)$$

where $\beta = 1/kT$ and the potential $U(\mathbf{r})$ per segment accounts for all interactions along the contour of the chain. A suitable definition of U as a "potential of mean force" (Freed, 1972), or, in other words, a mean field, decouples the many-body problem of segment interactions so that the chain propogates as a Markov process (van Kampen, 1981). Consequently, addition of another segment to a chain of $s - 1$ segments produces

$$G(\mathbf{r}, \mathbf{r}'; s) = \frac{1}{4\pi l^2} \int G(\mathbf{r} + l\mathbf{n}, \mathbf{r}'; s - 1) \exp\left[-\int_{s-1}^{s} \beta U[\mathbf{r}(s')] \, ds'\right] d\mathbf{n}, \qquad (28)$$

where $\mathbf{n}$ is the unit vector from segment s to $s - 1$. If U varies slowly on the scale of l, then

$$\int_{s-1}^{s} \beta U[\mathbf{r}(s)] \, ds \approx \beta U(\mathbf{r}) \qquad (29)$$

so that

$$G(\mathbf{r}, \mathbf{r}'; s) = \frac{e^{-\beta U(\mathbf{r})}}{4\pi l^2} \int G(\mathbf{r} + l\mathbf{n}, \mathbf{r}'; s - 1) \, d\mathbf{n}. \qquad (30)$$

Expanding $G(\mathbf{r} + l\mathbf{n}, \mathbf{r}'; s - 1)$ in a Taylor series in $\mathbf{r}$ and s, integrating, and rearranging gives the SCF equation

$$\frac{\partial G}{\partial s} = \frac{l^2}{6} \nabla^2 G + [1 - e^{\beta U(\mathbf{r})}] G, \qquad (31)$$

with the initial condition

$$G(\mathbf{r}, \mathbf{r}'; 0) = \delta(\mathbf{r} - \mathbf{r}'). \qquad (32)$$

Equation (31) describes phenomena (Chandrasekhar, 1943; Weiss and Rubin, 1983) ranging from the Brownian motion of colloidal particles to stellar evolution. For polymer molecules, the first term in Eq. (31) accounts for the connectivity of the chain, the second for the tendency of entropy to disperse the segments, and the third for all long-range interactions.

Since the chain statistics are governed by a Markov process, the chain may be divided into two independent subchains of lengths s and $n - s$. The probability of finding the sth segment at $\mathbf{r}$ for a chain that begins at $\mathbf{r}'$ and ends at $\mathbf{r}''$ is $G(\mathbf{r}, \mathbf{r}'; s)G(\mathbf{r}'', \mathbf{r}; n - s)/e^{-\beta U(\mathbf{r})}$; the denominator remedies the overcounting of the junction segment. Integrating over the endpoint positions and all contour locations where subchains may intersect yields

$$\varphi(\mathbf{r}) = \frac{e^{\beta U(\mathbf{r})}}{n} \int_0^n \iint G(\mathbf{r}, \mathbf{r}'; s) G(\mathbf{r}'', \mathbf{r}; n - s) \, d\mathbf{r}' \, d\mathbf{r}'' \, ds. \tag{33}$$

Once the SCF $U = U[\varphi(\mathbf{r})]$ is specified, Eqs. (31) and (33), and the necessary boundary, initial, and normalization conditions form a closed "self-consistent" set of equations.

Solutions for dilute ideal chains with $U \approx 0$ are particularly simple. Since segments do not interact, only internal configurations are important. With $\mathbf{r}' = 0$, Eq. (31) becomes

$$\frac{\partial G}{\partial s} = \frac{l^2}{6r^2} \frac{\partial}{\partial r}\left[r^2 \frac{\partial G}{\partial r}\right],$$

with $G(r, 0; 0) = \delta(r)$ and $G(\infty, 0; s) = 0$ as the boundary and initial conditions. The solution

$$G = \frac{3}{(2\pi l^2 s)^{3/2}} \exp\left[-\frac{3r^2}{2l^2 s}\right] \equiv G_0$$

indicates a Gaussian chain for which

$$\langle r^2 \rangle_0 = \frac{1}{W} \int r^2 G(r, 0, n) \, dr = nl^2,$$

as required.

The osmotic pressure calculation remains simple for all dilute solutions. Statistical mechanics gives the Helmholtz free-energy

$$A = -kT \ln(Q) \tag{34}$$

as a function of the canonical partition function Q. For $n_p \equiv \rho V/n$ chains in a volume V at high dilution,

$$Q \approx \frac{[VW]^{n_p}}{n_p!} \tag{35}$$

(McQuarrie, 1976). After using Stirling's approximation, Eqs. (34) and (35) yield

$$\frac{An}{\rho V k T} = \ln(\rho) - 1 - \ln(W) \tag{36}$$

as the free energy per chain. The free energy therefore depends on the internal state of the isolated chain, but the osmotic pressure

$$\frac{\Pi n}{\rho k T} = -\frac{Vn}{\rho k T}\left(\frac{\partial A}{\partial V}\right)_T = 1 \tag{37}$$

does not.

In concentrated solutions, U is a constant independent of position; Eq. (31) is again linear with the solution

$$G = G_0 \exp\{s[1 - \exp(\beta U)]\}, \tag{38}$$

demonstrating that chains are ideal. Between the dilute and concentrated limits, however, Q may not be factored into translational and internal parts (V and W) as in Eq. (34). Edwards' (1965) single-chain analysis is the foundation for all many-chain SCF theories. Expressions for Q take the form of functional integrals (Freed, 1972; Helfand, 1975c) which depend on many-body interactions and cannot be evaluated analytically.

Statistical mechanical manipulations of the functional integral representation of Q are necessary for inhomogeneous systems (Helfand, 1975c; Hong and Noolandi, 1981). Minimization of the free energy fixes the equilibrium spatial distribution of polymer and solvent. Edwards' "random field" technique (1965) leads to

$$U(\mathbf{r}) = \frac{\partial}{\partial \rho}[f(\rho, \rho_s) - \rho\mu_p^b - \rho_s\mu_s^b] - \frac{kT}{n}\ln(\rho/\rho_b) \equiv \Delta\mu_p, \tag{39}$$

where $\rho(\mathbf{r})$ and ρ_s are the segment and solvent densities, μ_p^b and μ_s^b are the respective chemical potentials in the bulk solution where $\rho \to \rho_b$, and $f(\rho, \rho_s)$ is the local free-energy density of a (perhaps hypothetical) mixture of composition ρ, ρ_s. Thermodynamically, $U(\mathbf{r})$ is the local excess chemical potential of the polymer; the logarithm term insures that U is invariant with respect to the choice of segment length (Hong and Noolandi, 1981). For a homogeneous solution, $U \to 0$ so that Eq. (39) provides no new information. Now the closed self-consistent set of Eqs. (31), (33), and (39) give the properties of a nonuniform solution in terms of the local segment interactions represented in $f(\rho, \rho_s)$. The segment equation-of-state must come from another source, such as the Flory–Huggins theory. Although its utility is limited in bulk solutions, the SCF method plays a central role in theories of surface tension and polymer adsorption.

III. Randomly Adsorbing Homopolymer

A. INTRODUCTION

In many systems of practical interest, polymer chains adsorb onto surfaces through contacts distributed randomly along the contour of the chain. These contacts result from attractive forces between polymer segments and the surface, generally attributed to van der Waals forces, hydrogen bonding, or other material-specific interactions. Measurements of polymer–surface interactions are relatively few in number (Killmann et al., 1977; Brebner et al., 1980a,b; Cohen Stuart et al., 1984a,b; Killmann and Bergmann, 1985, Van der Beek and Cohen Stuart, 1988). In spite of this, theories of random homopolymer adsorption have progressed remarkably by treating the energy of adsorption of a single segment as a phenomenological parameter. Random adsorption of individual segments and the concomitant impact on the overall configurations of adsorbed chains distinguished these systems from the related cases of terminally anchored chains and depletion layers created by nonadsorbed polymer.

Unlike terminally anchored polymer, both ends of a randomly adsorbed chain extend from the surface into solution; multiple interior points along the contour of the chain contact the surface. This architecture has generated its own terminology: Sequences of segments lying on the surface are "trains," nonadsorbed sequences with adsorbed segments at both ends are "loops," and sequences that leave the surface and never return are "tails." These structures are easily visualized and provide a conceptual link between the statistics of chain configurations and experimentally observable quantities. Furthermore, train, loop, and tail statistics measure the relative importance of competing physical phenomena: The loss of chain configurational entropy due to the impenetrable surface and interactions with other chains relative to energetic effects resulting from polymer–polymer, polymer–solvent, and polymer–surface interactions.

Such interactions control the structure of an adsorbed polymer layer. The equilibrium state of a system, including a surface, adsorbed and free polymer, and the solvent, of course, correspond to a minimum in the total free energy. The impenetrable adsorbing surface eliminates many of the configurations that the chain could assume in bulk solution. In addition, each adsorbed segment loses a translational degree of freedom. The resulting decrease in configurational entropy increases the total free energy and is unfavorable. Repulsion between segments on the same or different chains further reduces the number of available configurations and decreases the entropy. Despite this, adsorption still occurs when the energy decrease associated with segment–surface contacts compensates for the entropy loss. The balance between configurational entropy loss (due to the surface and interchain interactions) and energy gain (from adsorption) is central to the understanding of polymer adsorption.

The free energy is minimized to varying degrees by the formation of trains, loops, and tails, subject to the constraints of overall chain structure. For example, trains have low energies but also have low entropies; loops and tails effectively preserve entropy and promote segment–solvent contact but forego the adsorption energy at the surface. The polymer molecular weight and concentration in bulk solution and the solvent quality all affect the equilibrium amount of adsorbed polymer and the distribution of trains, loops, and tails within the adsorbed layer.

Recent models of polymer adsorption incorporate all these considerations, whereas early theories generally invoked one or more assumptions which limited their applicability. The most common approximations included the neglect of interchain interactions (i.e., isolated chain adsorption), restriction to infinite molecular weights, predetermination of the statistical distribution of loops and tails, and a priori specification of the polymer density profile. Nonetheless, the early theories provide the foundation for all succeeding work, treating the important physical phenomena individually and providing insights that might be overlooked today. Comprehensive reviews of early adsorption theories exist elsewhere (Stromberg, 1967; Silberberg, 1971; Vincent, 1974; Eirich, 1977; Tadros, 1982, 1985; Takahashi and Kawaguchi, 1982); we limit our treatment to those results that bear on contemporary analyses.

The development of theory for polymer adsorption parallels that for polymer solutions, since the two problems share the same physics (except for the effect of adsorption) on length scales comparable to the segment length. Several complementary approaches have emerged, including scaling theory, renormalization group theory, and self-consistent mean-field theory, to describe the layer structure in terms of parameters characterizing the universal aspects of polymer chains. Each idealizes the relevant physics, but the approximations and, hence, the ranges of validity differ as discussed later. In addition, recent experimental results will be presented and compared with theoretical predictions.

B. Early Adsorption Theories

1. *Individual Ideal Chains*

The development of models began in earnest (around 1950) when a growing body of experimental data suggested that macromolecular adsorption could not be described by concepts applied successfully to low molecular weight adsorbates. Jenckel and Rumbach (1951), for instance, observed adsorbed amounts corresponding to roughly 10 monolayers of segments. They reasoned that interactions were too weak in their dilute system to cause phase separation or multilayer adsorption; thus, the polymer must adsorb in alternating sequences of trains and loops. The first theoretical analysis based on this idea appeared in a series of papers by Simha, Frisch, and Eirich (Frisch et al., 1953; Simha et al., 1953; Frisch and Simha, 1954, 1956, 1957; Frisch, 1955). Polymer configurations are represented as ideal (Gaussian) random walks in a discrete three-dimensional lattice bounded by an impenetrable planar surface. Random walks reaching lattice sites on the surface are reflected back into the solution, a boundary condition motivated by the fundamental work of Chandrasekhar (1943).

However, Silberberg (1962b), DiMarzio (1965), and DiMarzio and McCrackin (1965) criticized the use of the reflecting boundary condition for overcounting allowable configurations. They argued in favor of the "adsorbing" boundary condition, i.e., discounting any configuration reaching a hypothetical lattice point *behind* the surface. Silberberg (1962a) also objected to decoupling the calculation of polymer configurations from the thermodynamic considerations which determine the equilibrium adsorbed amount and observed that "the discussion of the isolated macromolecule should be reformulated, and the shape of the polymer introduced as a variable into the full thermodynamic treatment."

Silberberg then addressed the equilibrium configuration of an isolated adsorbed macromolecule by dividing the chain into m_i loops of length i and $\tilde{m}_j$ trains of length j. The total number of segments in the chain is $n = \sum i m_i + \sum j \tilde{m}_j$, and, neglecting tails, $\sum m_i = \sum \tilde{m}_j$. This allows the partition function for the polymer to be expressed in terms of m_i, $\tilde{m}_j$, and the functions $\psi(i)$ and $\tilde{\psi}(j)$, specifying the number of possible configurations for loops and trains of lengths i and j, respectively. By assuming the loop and train size distributions to be sharply peaked at $i = P_B$ and $j = P_S$ [a serious restriction subsequently eliminated by Hoeve et al. (1965)] and choosing reasonable forms for $\psi(P_B)$ and $\tilde{\psi}(P_S)$, Silberberg minimized the free energy subject to the constraint of fixed chain length. The resulting simultaneous equations yielded P_B and P_S and the bound fraction of segments as $p = P_S/(P_B + P_S)$.

The configurational statistics derived from the model depend on the coordination numbers of the lattice and the energy of replacing an adsorbed solvent molecule with a segment, χ_s. The most notable result was a sharp transition from small p to $p \approx 1$ at low χ_s, the first indication of a critical energy χ_{sc} distinguishing depletion from adsorption. Silberberg (1962b) concluded that polymer random walks "can continue in [the surface] in two out of three dimensions, and that a small reduction in energy per segment can compensate largely for the entropy loss in the third."

Hoeve et al. (1965) carried this concept forward by varying the lattice geometry to account for chain flexibility and using correct combinatorial factors in the partition function with no a priori assumptions regarding the size distributions of loops and trains. In the limit of very stiff chains, the bound fraction becomes discontinuous: $p = 0$ for $\chi_s < \chi_{sc}$ and $p = 1$ for $\chi_s > \chi_{sc}$. Some flexibility broadens the transition, but a critical adsorption energy always exists below which $p = 0$. They also found (1) broad train and loop size distributions with no maxima, (2) average loop lengths that increase with decreasing flexibility and adsorption energy and become infinite at the critical energy (for infinite chain lengths), and (3) average train lengths that increase with increasing flexibility and adsorption energy. Later work by Silberberg (1967) confirmed these results.

Roe (1965, 1966) and Rubin (1965) completed the description of the dependence of bound fraction and average loop and train sizes on adsorption energy, chain flexibility, and molecular weight for isolated macromolecules. Clear distinctions emerged among adsorption energies greater than, equal to, or less than the critical energy. For $\chi_s < \chi_{sc}$, $v = np$ (the number of adsorbed segments per chain) and is independent of the chain length n; at $\chi_s = \chi_{sc}$, $v \simeq n^{1/2}$ (also shown in Simha et al., 1953); and for $\chi_s > \chi_{sc}$, $v \simeq n$, implying a constant bound fraction. The bound fraction never falls to zero for finite chain lengths, but a well-defined transition, dependent on the lattice geometry or the chain flexibility, still exists at the critical energy. The variation of the average loop and train lengths with adsorption energy are the same as found earlier by Hoeve et al. (1965). Although the train length does not vary significantly with chain length, the loop length does: For $\chi_s < \chi_{sc}$, loop length is independent of n for $n > 100$, but for $\chi_s > \chi_{sc}$, loop length is proportional to $n^{1/2}$.

Roe's analysis (1965), the first to allow for tails, is highly significant. First, the average tail length decreases with stronger adsorption, as expected. For $n > 1000$ segments, the tail length exceeds the loop length by an order of magnitude and increases in proportion to the chain length. Below the critical adsorption energy, the root–mean–square distance of segments from the surface is proportional to the radius of gyration, $(nl^2/6)^{1/2}$; above the critical energy, the thickness is only a few segments. Loops dominate the segment density profiles for $\chi_s > \chi_{sc}$, while tails are more important for $\chi_s \leq \chi_{sc}$. In fact, near the critical energy, about 70% of all segments are in tails and "the thickness of [the adsorbed] layer ... is largely determined by the size of the dangling polymer chain ends, rather than by the size of the loop, as was frequently assumed."

This conclusion can easily be rationalized. Chandrasekhar (1943; Hoeve, 1965; Hesselink, 1969) has shown that for long tails of i segments, the number of possible configurations is $(2\pi i)^{-1/2} 2^i$, while the number of possible loop configurations is $(2\pi i)^{-1/2} 2^i i^{-1}$. Thus, configurational entropy alone favors tails over loops. Near the critical adsorption energy, the attractive surface binds the chain but does not distort it significantly from its ideal state. Thus, tails dominate. Stronger adsorption overcomes the entropic effects and flattens the chain onto the surface. Then tails can be neglected for isolated macromolecules without excluded volume effects.

2. *Equilibrium and Excluded Volume*

In practice, macromolecules almost never adsorb in isolation. Although the adsorption energy per segment may be low, the number of adsorbing segments per chain results in the deposition of many chains, even from dilute bulk solutions. Hence, two intimately related problems must be addressed: Determination of the equilibrium adsorption as a function of bulk solution concentration (the adsorption isotherm) and assessment of the effects of excluded volume and solvency. Silberberg (1962b) laid the foundation for subsequent developments by demonstrating two key concepts: (1) The adsorption energy equals the energy required to replace an adsorbed solvent molecule with a segment and (2) a mixing parameter that includes entropic and energetic contributions adequately characterizes solvent–segment contacts.

Simha *et al.* (1953), Silberberg (1962b, 1968), and Hoeve, *et al.* (1965; Hoeve, 1966) addressed the problem of equilibrium, predicting adsorption isotherms in at least qualitative agreement with experimental data. On a linear scale, the adsorbed amount grows very steeply with polymer volume fraction and then levels off in a "pseudoplateau." The adsorbed amount increases with chain length, reaching a limit of about one monolayer of segments for long chains adsorbed from dilute solutions.

In retrospect, the early adsorption models identified most of the significant aspects of the adsorption problem. The relationship between adsorption energy and average chain configuration was explored extensively. The assumptions necessary for the calculations, however, limited the utility of these models in predicting the behavior of real systems and, hence, their credibility. Improved analyses of excluded volume and equilibrium required a more comprehensive mathematical framework.

C. Lattice Models and Matrix Methods

Lattice models play a central role in the description of polymer solutions as well as adsorbed polymer layers. All of the adsorption models reviewed so far assume a one-to-one correspondence between lattice random-walks and polymer configurations. In particular, the general scheme was to postulate the train–loop or train–loop–tail architecture, formulate the partition function, and then calculate the equilibrium statistics, e.g., bound fraction, average loop and train sizes, and segment density distribution (Roe, 1966; Hoeve, 1965). In some cases, the adsorbed macromolecules were equilibrated with bulk solution, but the analyses usually focussed on isolated chains with at least one segment on the surface.

The work of DiMarzio and Rubin (DiMarzio, 1965; Rubin, 1965; DiMarzio and Rubin, 1971) began the development of a related but more powerful approach. Rather than calculating microstructural details from a presumed architecture, Rubin's matrix method concentrates on the effect of local interactions on the propagation of the chain, thereby deriving the statistical properties of the random walk and the structure of the entire chain. This formalism is the foundation for several subsequent models, so some details are reviewed here. The notation is transposed into a form consistent with the contemporary models discussed below.

The space near the adsorbing surface is discretized into layers of lattice sites numbered $i = 1, 2, \ldots m$ with layers 1 and m being adsorbing planar surfaces and $m \to \infty$ corresponding to two isolated surfaces. Each lattice site has Z nearest neighbors with Z_0 of these in the same layer. Far from the surface, all random walks have the same *a priori* probability, but adsorption changes the statistical weights of configurations near the surface.

Let $P(i, s)$ be the unnormalized probability that the sth step of a polymer random walk ($s + 1$ segments) ends at a lattice site in layer i. The preceeding step $s - 1$ must reside in one of the layers $i - 1$, i, or $i + 1$, suggesting the recurrence relation

$$P(1, s) = \lambda_0 \exp(\chi_s) P(2, s - 1) + \lambda_1 \exp(\chi_s) P(1, s - 1), \qquad i = 1$$

$$P(i, s) = \lambda_1 P(i + 1, s - 1) + \lambda_0 P(i, s - 1) + \lambda_1 P(i - 1, s - 1), \quad i \geq 2 \quad (40)$$

where $\lambda_1 = \frac{1}{2}(Z - Z_0)/Z$ and $\lambda_0 = Z_0/Z$ are the fractions of nearest neighbors in adjacent layers or the same layer, respectively. The term $\exp(\chi_s)$ in Eq. (40) enhances the probability of configurations contacting the surface. In matrix form, the equation

$$\mathbf{P}(s) = \mathbf{W}(\chi_s)\mathbf{P}(s - 1) = [\mathbf{W}(\chi_s)]^s \mathbf{P}(0) \tag{41}$$

relates $\mathbf{P}(s)$, the column vector with elements $P(i, s)$, $i = 1, \ldots, \infty$, to $\mathbf{P}(0)$, the vector of starting probabilities $[\mathbf{P}(0)]^T \equiv [e^{\chi_s}, 1, 1, \ldots]$, through $\mathbf{W}$, a matrix of transition probabilities

$$\mathbf{W} \equiv \begin{bmatrix} \lambda_0 e^{\chi_s} & \lambda_1 e^{\chi_s} & & \\ \lambda_1 & \lambda_0 & \lambda_1 & \\ & \lambda_1 & \lambda_0 & \lambda_1 & \cdots \\ & & \vdots & & \end{bmatrix}. \tag{42}$$

Structural characteristics of adsorbed chains, such as the mean distance of chain ends from the surface and the fraction of segments in each layer, derive from $P(i, s)$. The partition function and thermodynamic properties depend on the eigenvalues of $\mathbf{W}$ (Flory, 1969). In the limit $n \to \infty$, calculation of thermodynamic functions simplifies because one eigenvalue, denoted by Λ, dominates the free energy per segment,

$$a = -kT \ln(\Lambda), \tag{43}$$

and the entropy per segment,

$$s = k \frac{\partial}{\partial \chi_s} \left[\chi_s \ln(\Lambda) \right]. \tag{44}$$

Locating a second adsorbing surface at layer $i = m$ indicates $a = a(m)$; the force between the surfaces

$$-\frac{da(m)}{dm} = a(m) - a(m+1) = -kT \ln \left[\frac{\Lambda(m)}{\Lambda(m+1)} \right] \tag{45}$$

is negative for attraction.

DiMarzio and Rubin's predictions (Rubin, 1965; DiMarzio and Rubin, 1971) confirm earlier results (Hoeve et al., 1965; Silberberg, 1967; Roe, 1965, 1966) and provide new ones. The entropy per segment is zero for $\chi_s < \chi_{sc}$, but decreases considerably as chains lose configurational freedom upon adsorption for $\chi_s > \chi_{sc}$. The force between surfaces is monotonically repulsive for $\chi_s < \chi_{sc}$ and increases as χ_s decreases. For $\chi_s > \chi_{sc}$, monotonic attraction is predicted, although its strength at fixed separation passes through a maximum and then falls as χ_s increases. These predictions, however, do not allow for exchange of polymer with the bulk solution.

The most significant aspect of this work is the opportunity for extension afforded by the matrix formalism. DiMarzio (1965) and Rubin (1965) suggested incorporating excluded volume by multiplying each row of $\mathbf{W}$ by a factor $e^{\beta U(i)}$ where $U(i)$, the potential energy of segments in layer i, includes the energy of segment–segment, segment–solvent, and segment–surface interactions. In turn, the spatial distributions of these interactions depend on the overall configuration of chains in the adsorbed layer to be calculated from $\mathbf{W}$. The coupling between the configuration probability and the hypothetical potential field is the central idea in contemporary SCF models discussed in the next section.

Other lattice models are noteworthy as well. Roe (1974), for instance, developed a statistical mechanical formulation for an adsorbed layer capable of exchanging polymer and solvent with the bulk solution. The grand canonical ensemble, first introduced by DiMarzio and Rubin (1971), ensures an equilibrium distribution of polymer and solvent. Segment–solvent contacts, each having an energy characterized by the Flory–Huggins parameter χ, are counted within a mean-field approximation based on a random distribution of segments and solvent in each layer. The purely configurational part of the partition function closely follows Flory's derivation of the entropy of mixing polymer with solvent in bulk solutions, thereby consistently describing excluded volume in adsorbed layers and in bulk solutions.

In addition to the assumptions inherent in the Flory–Huggins theory, Roe and others introduced another approximation. In general, the distribution of segments depends on spatial position as well as contour location or rank within the chain. Roe (1974) neglected the latter dependence, making all segments equivalent to middle segments and precluding the treatment of tails and other end effects. This approximation was obscured by Roe's focus on volume fractions rather than chain configurations. The models of Helfand (1975a, 1976; Weber and Helfand, 1976) for bulk polymer interfaces rely directly on statistical analysis of configuration probability while incorporating the Flory–Huggins theory for chain entropy. Likewise, Levine et al. (1978) treated solvency and excluded volume by analogy with polymer-solution theory within the matrix formalism of DiMarzio and Rubin (1971). These efforts also ignore the influence of segmental rank within the chain on the spatial distribution of segments, a limitation overcome in contemporary adsorption models discussed later.

D. Scaling and Renormalization Group Theories

1. *Self-Similarity*

Several complementary approaches, roughly paralleling those for polymer solutions, have emerged for modelling polymer adsorption. The introduction of scaling concepts (de Gennes, 1981) has stimulated considerable discussion (Fleer et al., 1988) concerning the applicability of this method relative to traditional mean-field theories. Renormalization group theory may help resolve this question, but the technique has only recently yielded results (Eisenriegler, 1985; Nemirovsky and Freed, 1985).

The scaling analysis of long flexible chains in a good solvent in contact with a surface (de Gennes, 1981; for reviews see Cohen Stuart et al., 1986; de Gennes, 1987; Fleer et al., 1988) relies heavily on the solution results introduced earlier. An adsorbed layer can be pictured as a transient network of entangled chains (de Gennes, 1981) with correlation length ξ which varies with z, the distance from the surface. Since ξ and z are the only length scales in the problem, dimensional analysis requires that $\xi \approx z$, making the adsorbed layer "self-similar." Thus, the volume fraction must satisfy $z \approx \xi[\varphi(z)] \approx l\varphi^{-3/4}$ so that

$$\varphi(z) \approx \left[\frac{z}{l}\right]^{-4/3} \quad (46)$$

results as the fundamental scaling prediction. Note that the functional form of Eq. (46) is independent of the segment–surface interaction potential.

Of course, the self-similarity of the layer has limits. The correlation length cannot exceed that in bulk solution; as z increases, $\xi \to \xi_b \approx l\varphi_b^{-3/4}$. Distances $z > \xi_b$ thus define the "distal" region where $\varphi(z)$ decays exponentially (de Gennes, 1981). The lower bound on Eq. (46) is set by a length D that depends on the adsorption energy. For $z < D$ (defining the "proximal" region), the adsorption energy influences $\varphi(z)$. Scaling analyses for the proximal region have appeared (Eisenriegler et al., 1982; Eisenriegler, 1983; de Gennes and Pincus, 1983). At such small length scales, surface roughness or molecular effects may also enter (Cohen Stuart et al., 1986) when adsorption is strong. Universal behavior, though, is expected in the "central" region $D < z < \xi_b$ governed by Eq. (46).

2. Adsorbed Amount and Free Energy

The remaining problem is to determine D or, alternately, the apparent surface volume fraction φ_s. De Gennes (1981) originally conjectured that $D \approx \xi(\varphi_s) \approx l\varphi_s^{-3/4}$, but subsequent work by Eisenriegler et al. (1982; Eisenriegler, 1983) leads to $D \approx l\varphi_s^{-1}$ (de Gennes and Pincus, 1983) as supported by Monte Carlo simulations and exact enumerations (Ishinabe, 1982). D (or φ_s) may now be fixed by minimizing the layer's free energy.

The free-energy density of the inhomogeneous system may be expanded in powers of $\nabla\varphi, \nabla^2\varphi, \ldots$, with the first term representing the free energy needed to create a unit volume of solution at φ from a reservoir at φ_b (Cahn and Hilliard, 1958; Cahn, 1977; Widom, 1979). Integration over z gives the surface tension γ (free energy per unit area of interface) in the form (Cahn, 1977)

$$\gamma = \gamma_d(\varphi_s) + I[\varphi_s, \varphi_b] \quad (47)$$

where γ_d accounts for segment–surface interactions and I is an integral measure of the energy penalty associated with maintaining the composition gradient of the layer. Next, γ_d is expanded as

$$\gamma_d = \gamma_0 + \gamma_1 \varphi_s + O(\varphi_s^2) \approx \gamma_0 + \gamma_1 \left[\frac{D}{l}\right]^{-1}, \quad (48)$$

with $\gamma_1 < 0$ for adsorption; truncation at $O(\varphi_s)$ assumes $|\gamma_1| l^2 / kT = \chi_s \ll 1$, i.e., weak segment–surface interactions. Since contributions to the free energy due to the composition gradient are greatest near the surface, de Gennes (1981; de Gennes and Pincus, 1983) assumed $I[\varphi_s, \varphi_b] \approx I(\varphi_s)$ and, with Eq. (12), integrated over a thickness comparable to D to obtain

$$I(\varphi_s) \approx \int_0^D \frac{kT}{\xi^3(\varphi_s)} dz \approx \frac{kT}{D^2}. \tag{49}$$

Combining Eqs. (47)–(49) and minimizing with respect to D yields

$$\frac{D}{l} \approx \left[\frac{|\gamma_1|l^2}{kT}\right]^{-1} = \chi_s^{-1}, \tag{50}$$

indicating that $D > l$ for $\chi_s < 1$.

Integrating Eq. (46) gives the adsorbed amount of polymer as

$$\Gamma \equiv \int_0^\infty \frac{\varphi(z)}{l^3} dz \approx \frac{1}{l^2}\left[\frac{l}{D}\right]^{1/3} \approx \frac{1}{l^2}\left[\frac{|\gamma_1|l^2}{kT}\right]^{1/3} \tag{51}$$

in units of monolayers (i.e., number of segments per segment area). For γ_1 not too small, Γ is $O(1)$ and independent of φ_b and n; mean-field theories predict qualitatively similar behavior for $v > 0$ and $n \to \infty$, where scaling is valid.

A subsequent analysis (de Gennes, 1982) provides the scaling form of the interaction potential between planar adsorbed layers at separation h and highlights the question of equilibrium. For $h \gg D$ and full equilibrium, i.e., polymer in the gap freely exchanging with the bulk solution,

$$\varphi(h/2) \approx \left[\frac{l}{h}\right]^{4/3}$$

and

$$\Pi(h/2) \approx \frac{kT\varphi^{9/4}}{l^3} \approx \frac{kT}{h^3}$$

from Eq. (46) and Eq. (12). Variations of Eq. (47), $\delta\gamma/\delta h$, include contributions from the change of integration limits in $I[\varphi_s, \varphi_b]$ and from variations in the profile $\varphi(z)$. Neglecting the latter, de Gennes finds an attractive interaction potential

$$\frac{\Delta A(h)}{l^2} \approx -\int_h^\infty (h'/2) dh' \approx \frac{-kT}{h^2}.$$

On the other hand, holding the adsorbed amount Γ constant introduces a constraint that is incorporated in the minimization of γ through a Lagrange multiplier. Although the details of the original analysis (de Gennes, 1982) are not completely correct (de Gennes and Pincus, 1983), the resulting repulsion $\Delta A(h)/l^2 \approx kT/h^2$ is universal for $h \gg D$ and this "restricted" equilibrium condition that allows relaxation of chains within the layer but no exchange with the bulk solution. Scaling analyses do not predict attraction and repulsion in the same potential curve.

3. *Comparison with Scattering Measurements*

Neutron-scattering experiments allow a direct test of Eq. (46) for polymer layers on an adsorbent having a high surface area-to-volume (S/V) ratio. The tendencies of the adsorbent, solvent, and polymer to scatter neutrons (also known as the "contrast" relative to background noise) are adjusted through the isotopic compositions of the components. Scattering theory (Cebula *et al.*, 1978; Auvray and Cotton, 1987; Cosgrove *et al.*, 1987a) relates the measured intensity to "structure factors" which characterize the density profiles in the system.

Matching the contrasts of the adsorbent and solvent enables detection of the structure factor S_{pp} for polymer–polymer correlations within the adsorbed layer, defined by

$$S_{pp}(q) \equiv 2\pi \left[\frac{S}{V}\right] q^{-2} \left| \int_0^\infty \langle \varphi(z) \rangle e^{iqz} \, dz \right|^2 + \tilde{S}_{pp}, \qquad (52)$$

where $q = 4\pi/\lambda \sin(\theta/2)$, λ is the neutron wavelength, θ is the scattering angle, $\langle \varphi \rangle$ denotes the average of φ over directions parallel to the surface, and $\tilde{S}_{pp}$ is a contribution due to fluctuations of φ about $\langle \varphi \rangle$. The data of Auvray and Cotton (1987) (Fig. 9), for poly(dimethylsiloxane) (PDMS) adsorbing on silica

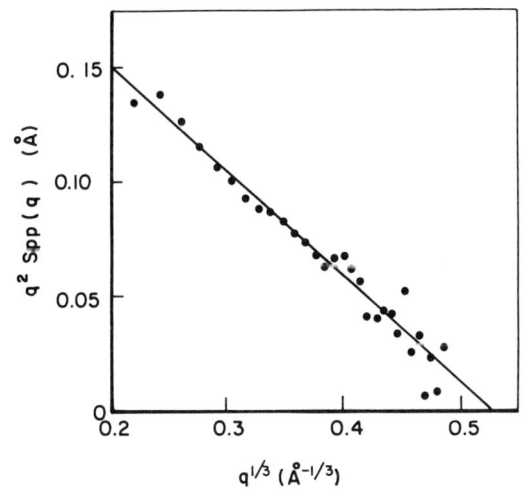

FIG. 9. Structure factor for polymer–polymer correlations plotted as $q^2 S_{pp}(q)$ vs. $q^{1/3}$, for PDMS adsorbing on silica from cyclohexane (Auvray and Cotton, 1987).

in cyclohexane, confirms the scaling prediction

$$S_{pp}(q) \approx (ql)^{1/3}, \tag{53}$$

derived from Eqs. (46) and (52) by Auvray and de Gennes (1986). Concentration fluctuations, which produce $\tilde{S}_{pp} \approx (ql)^{2/3}$, are not observed on the large length scales (small q values) probed by the experiment.

Adjusting the contrast difference between the adsorbent and the solvent away from zero (Cebula *et al.*, 1978) reveals the structure factor

$$S_{pg}(q) = 2\pi \left[\frac{S}{V}\right] q^{-3} \int_0^\infty \langle \varphi(z) \rangle \sin(qz)\, dz \tag{54}$$

for adsorbent–polymer correlations. Combining Eq. (54) and Eq. (46) yields

$$S_{pg}(q) \approx (ql)^{-8/3} \tag{55}$$

as the scaling prediction; Fig. 10 hows the corresponding data of Auvray and Cotton (1987). For several reasons, unambiguous measurement of S_{pg} is difficult and entails considerable experimental error (Auvray and Cotton,

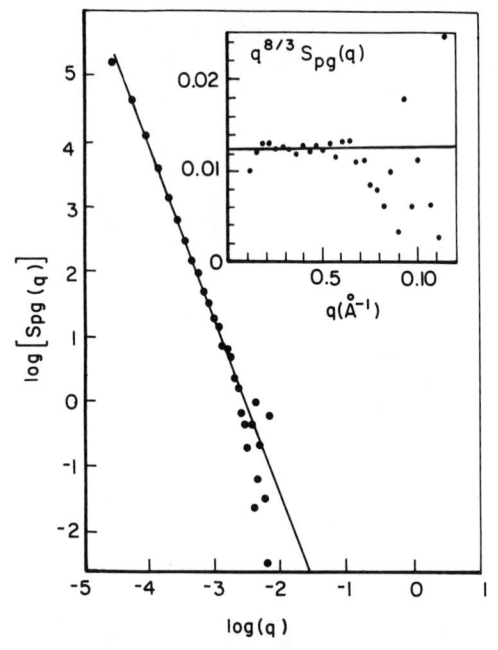

FIG. 10. Structure factor for polymer–adsorbent correlations plotted on a log–log scale, and as $q^{8/3}S_{pg}(q)$ vs. q in the inset, for PDMS adsorbing on silica from cyclohexane (Auvray and Cotton, 1987).

1987; Cosgrove *et al.*, 1987b); still, the data apparently exhibit the $q^{-8/3}$ trend.

Other neutron-scattering data, for poly(ethylene oxide) (PEO) adsorbing on deuterated polystyrene latex in water (Cosgrove *et al.*, 1987a,b), clearly *do not* show scaling behavior. Schaefer's (1984) analysis indicates that solutions of very flexible polymers, such as PDMS, have a large semidilute-good regime in which scaling is valid, whereas solutions of slightly stiffer and much shorter chains, such as PEO, are more aptly modelled by mean-field theory for moderate φ. Thus the T-c diagrams, though qualitative, do suggest a rationalization of these two sets of otherwise conflicting polymer adsorption data.

4. *Renormalization Group Approaches*

Recent efforts have capitalized on analogies with other critical phenomena in applying RG methods to polymer adsorption and depletion. Eisenriegler *et al.* (1982) consider isolated chains in both theta and good solvents but only for adsorption energies near the critical value. Eisenriegler (1983, 1984, 1985) extended the analysis to semidilute solutions contacting a surface with an arbitrary affinity for segments. Explicit functional forms for $\varphi(z)$, $\varphi(0)$, and Γ emerge, although the crossover behavior between theta and good solvent conditions is not detailed. In general, Eisenriegler's results verify the scaling predictions, most notably Eq. (46), and introduce several new predictions including the scaling form of the surface tension.

Freed's criticism (1987) of the magnet analogy, which also applies to scaling theory, concerns the inability of these methods to predict the molecular weight dependence of properties such as the mean layer thickness. Nemirovsky and Freed (1985) calculated layer properties via the "conformation–space" RG method developed for polymer solutions (Freed, 1987). The chain partition function and moments of the vector between chain ends, evaluated for theta and good solvent conditions, provide the crossover behavior for surface affinities ranging from strong repulsion to moderate attraction. However, the complexity of the analysis, especially the notation, makes assessment of the results difficult. Other efforts (Douglas *et al.*, 1986, 1987) are primarily concerned with polymer interacting with penetrable surfaces.

E. THE SCHEUTJENS-FLEER LATTICE MODEL

1. *Motivation and Formulation*

Each of the lattice models reviewed in the previous section introduces important concepts but fails to describe adsorbed layers satisfactorily over a wide range of conditions. The most complete synthesis to date is the model of Scheutjens and Fleer (SF) (Scheutjens and Fleer, 1979, 1980, 1985; Fleer and Scheutjens, 1986; see also the discussion following Levine et al., 1978). The SF model implements DiMarzio and Rubin's (1971) suggestion that the probability of each step in the chain's random walk should depend on a local potential field which accounts for all interactions. This mean field is averaged parallel to the surface and neglects correlations between segments. Like Roe's model (1974), the SF model resides in the grand canonical ensemble, permitting the exchange of polymer and solvent with bulk solution. Their free energy minimization with respect to a function measuring the probability of chain configurations is equivalent to Helfand's (1975c) method for establishing the "self-consistent" mean field.

The many reviews of the SF model (Tadros, 1982, 1985; Takahashi and Kawaguchi, 1982; Fleer and Lyklema, 1983; Cohen Stuart et al., 1986; Fleer et al., 1988; Fleer, 1988) reflect both the importance of polymer adsorption and the ability of the SF model to predict many features of adsorbed layers for wide ranges of polymer molecular weight, bulk solution concentration, solvent quality, and adsorption energy.

As in earlier lattice models, the solution is divided into layers of lattice sites parallel to the surface, labeled $i = 1, 2, \ldots$, having fractions λ_1 of nearest neighbors in each adjacent layer and $\lambda_0 = 1 - \lambda_1$ in the same layer, respectively. As in Flory–Huggins solution theory (see section II.D), all sites are occupied by either a solvent molecule or a chain segment, and every chain contains r sequential segments numbered $s = 1, 2, \ldots, r$. A function $P(i, s)$ specifies the unnormalized probability that any subchain of $s \leq r$ contiguous segments ends in layer i. In the bulk, all configurations are equivalent so that $P(i, s) \to 1$ as $i \to \infty$. The "initial" condition $P(i, 1)$ represents the probability of finding a subchain of length 1, i.e., individual segments, in layer i. Following the development of DiMarzio and Rubin (1971), $P(i, s)$ is constructed recursively from

$$P(i, s) = P(i, 1)[\lambda_1 P(i - 1, s - 1) + \lambda_0 P(i, s - 1) + \lambda_1 P(i + 1, s - 1)] \quad (56)$$

beginning with $P(i, 1)$. The term in brackets is the probability that a subchain of $s - 1$ segments ends at a site adjacent to the sth segment; multiplication by $P(i, 1)$ accounts for the likelihood of finding a segment at the site of interest.

Although DiMarzio and Rubin (1971) assumed $P(i, 1) \equiv 1$ for $i > 1$, the actual value depends on the mean field $U(\varphi_i)$ through

$$P(i, 1) \equiv e^{-\beta[U(\varphi_i) - U(\varphi_b)]} \quad (57)$$

where φ_i and φ_b indicate the segment volume fractions in layer i and in bulk solution. The appropriate form of $U(\varphi_i)$ minimizes the free energy of the layer and, therefore, produces the equilibrium distribution of chain configurations.

The form derived by Scheutjens and Fleer (1979) depends on the Flory–Huggins expression for the free energy of a solution [Eq. (22)]. The Helfand (1975b) SCF [Eq. (39)] can incorporate any equation-of-state; indeed, supplementing Eq. (22) by a term accounting for adsorption energy and substituting into Eq. (39) produces

$$\beta U(\varphi_i) = -\ln(1 - \varphi_i) - 2\chi\langle\varphi_i\rangle - \delta_{i,1}(\chi_s + \lambda_1\chi), \tag{58}$$

as derived by Scheutjens and Fleer (1979). Note that $\chi_s > 0$ for adsorption and $\delta_{i,1} = 1$ if $i = 1$, and $\delta_{i,1} = 0$ if $i > 1$;

$$\langle\varphi_i\rangle \equiv \lambda_1\varphi_{i-1} + \lambda_0\varphi_i + \lambda_1\varphi_{i+1} \tag{59}$$

accounts for nonlocal segment–solvent interactions. Equation (57) then provides

$$P(i, 1) = \frac{1 - \varphi_i}{1 - \varphi_b}\exp[\delta_{i,1}(\chi_s + \lambda_1\chi) + 2\chi(\langle\varphi_i\rangle - \varphi_b)] \tag{60}$$

as the initial condition for Eq (56).

Now, the system of equations must be closed by relating φ_i to $P(i, s)$. Since every segment within a chain of total length r also belongs to two "subchains" of lengths s and $r - s + 1$, the volume fraction is proportional to the joint probability of the two subchains meeting in layer i summed over all contour locations within the chain,

$$\varphi_i = \frac{\varphi_b}{P(i, 1)r}\sum_{s=1}^{r} P(i, s)P(i, r - s + 1), \tag{61}$$

where the denominator eliminates the double counting of the junction segment. Note that as $i \to \infty$, $\varphi_i \to \varphi_b$ as required.

Given the lattice parameters λ_1, λ_0 and the physical parameters r, φ_b, χ, and χ_s, Eqs. (56), (60), and (61) form a closed system that, when solved numerically, completely describes an adsorbed polymer layer. From φ_i, simple calculations give the adsorbed amount

$$\varphi_{ads} = \sum_{i=1}^{\infty} [\varphi_i - \varphi_b], \tag{62}$$

the surface coverage φ_1, the bound fraction $p = \varphi_1/\varphi_{ads}$, and higher moments of φ_i corresponding to various integral measures of the layer thicknesses. The statistics of trains, loops, and tails (including their relative numbers, length distributions, etc.) follow from $P(i, s)$ with minor computational effort. Evaluation of the partition function produces the surface tension of the layer. Calculation of the hydrodynamic thickness (Cohen Stuart et al., 1984c) requires an additional constitutive relation for the permeability of the adsorbed layer as a function of φ_i plus solution of the equations governing creeping flow. Placement of a second adsorbing surface at $i = m$ allows the formation of bridge sequences and alters the distribution of trains, loops, and tails. The interaction potential between the surfaces simply equals the difference between the surface tensions at separations of m and $m = \infty$.

2. Structure of Adsorbed Layers

Extensive results have been presented in the numerous papers and reviews cited earlier. In general, φ_i decreases steeply within a few layers near the surface followed by a gradual asymptote to φ_b. Fig. 11 (Scheutjens and Fleer, 1980) shows a typical plot of $\log(\varphi_i)$ vs. i, which highlights interesting features away from the surface. Apparently, φ_i decays as the sum of two exponentials with characteristic decay lengths that differ for low and high i. Segments in loops dominate near the surface but become less important as i increases. The density of tail segments reaches a maximum near the surface and dominates the profile at large i. Segments in nonadsorbed chains (not shown) are depleted near the surface; $\varphi_i \to \varphi_b$ at roughly $i \approx r^{1/2}$. Comparisons of predicted profiles and those obtained through neutron scattering (Cosgrove et al., 1987b) indicate qualitative agreement of the general features.

The distribution of segments in trains, loops, and tails depends on all of the experimental variables. Typically, as χ_s increases, the fraction of segments in trains increases while the fractions in loops and tails decrease. The opposite trends are noted as φ_b and r increase, although the competition *between* loops and tails is more complex. For example, Fig. 12 (Cohen Stuart, 1980) shows that as r grows, most segments reside in loops and tails, although the tail fraction falls slightly at high r. In most cases of practical interest, from 10% to

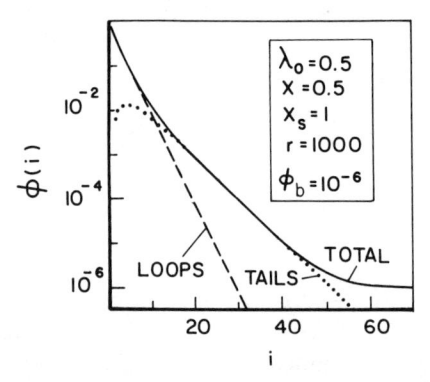

FIG. 11. Polymer volume-fraction profiles calculated by Scheutjens and Fleer (1980) and plotted as $\log(\varphi_i)$ vs. i. Parameters are given in the inset.

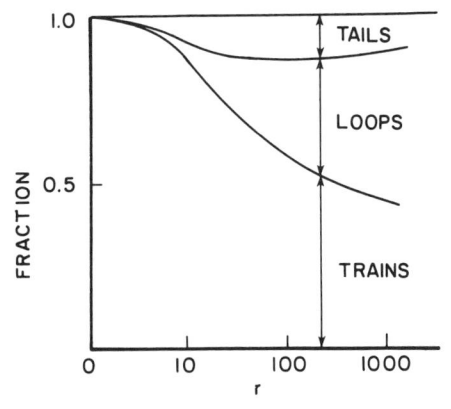

FIG. 12. Percentages of segments contained in trains, loops, and tails as a function of chain length, r, for a typical (but unspecified) set of parameters (Cohen Stuart, 1980).

25% of the segments in adsorbed chains are in tails; this value can rise as high as 67% for bulk polymer in contact with a surface.

At low values of φ_b, adsorption isotherms (φ_{ads} vs. φ_b) from the SF model rise steeply to an apparent plateau. A gradual rise in φ_{ads} continues as $\log(\varphi_b)$ increases over several decades. Most experimental adsorption isotherms for monodisperse polymer (Takahashi and Kawaguchi, 1982; Cohen Stuart et al., 1986) exhibit the same trends.

Several reviews (e.g., Takahashi and Kawaguchi, 1982; Cohen Stuart et al., 1986) discuss the extensive experimental data and the predictions of the SF model for the variation of φ_{ads} with polymer molecular weight. In general φ_{ads} increases linearly with $\log(r)$ for adsorption from a theta solvent, but is less in a good solvent, and appears to be limited at high r. Proper selection of parameters in the model produces excellent agreement with experimental data (Fig. 13) for both φ_{ads} and p.

Two measures of the layer thickness have been widely reported (Takahashi and Kawaguchi, 1982; Cohen Stuart et al., 1986). The root-mean-square layer thickness, defined through

$$t_{rms}^2 \equiv \frac{1}{\varphi_{ads}} \sum_{k=0}^{\infty} i^2 \varphi_i, \qquad (63)$$

can be compared with the thickness measured by ellipsometry. For theta solvents, Scheutjens and Fleer (1979) find $t_{rms} \approx r^{1/2}$, in agreement with other theories (Section II,C) and considerable experimental data. Experiments reveal $t_{rms} \approx r^{0.4}$ in good solvents, in accordance with the theory, but indicate

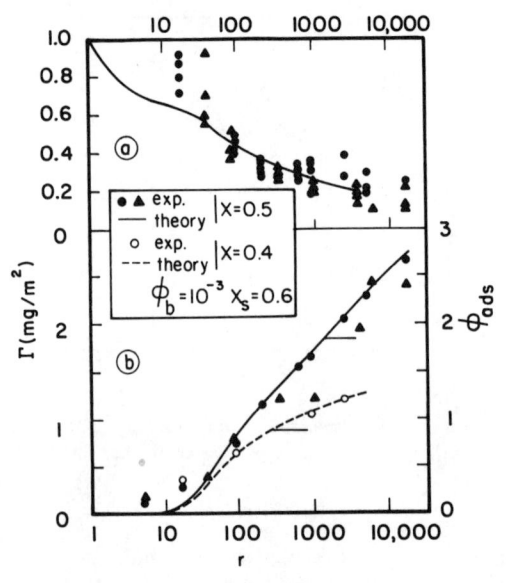

FIG. 13. Comparison of measured and calculated bound fraction (a) and adsorbed amount (b) as functions of chain length, r. The left ordinate in (b) applies to experimental points: Full circles and open circles denote PS/silica/cyclohexane and PS/silica/CCl$_4$ (vander Linden and van Leemput, 1978); triangles denote PS/silica/cyclohexane (Kawaguchi et al., 1980). The right ordinate applies to values calculated by Fleer and Scheutjens (1982a) for $\lambda_0 = 0.5$, $\chi = 0.50$ (full curves) and $\chi = 0.40$ (dashed curve); other parameters are given in the inset.

greater layer thicknesses than in a theta solvent, though the SF model predicts the opposite.

The hydrodynamic thickness, t_{hd}, is most significant for colloidal particles bearing adsorbed polymer. Theoretical calculations (Cohen Stuart et al., 1984c; Anderson and Kim, 1987) model the adsorbed layer as a porous medium with spatially-varying permeability and assume that the flow of solvent through the adsorbed layer does not perturb the layer structure. The predictions of the SF model and data for the PEO–polystyrene latex–water system (Figs. 14 and 15) (Cohen Stuart et al., 1984c) agree very well and indicate $t_{hd} \approx r^{0.8}$. The results vary little with χ and χ_s but require specification of the hydrodynamic drag associated with each segment. Reasonable values (e.g., a segmental drag equal to that of a sphere of comparable size) produce qualitatively correct behavior.

3. *Interactions Between Layers*

The interaction between adsorbed polymer layers is an important component of the total potential between colloidal particles. For collisions between

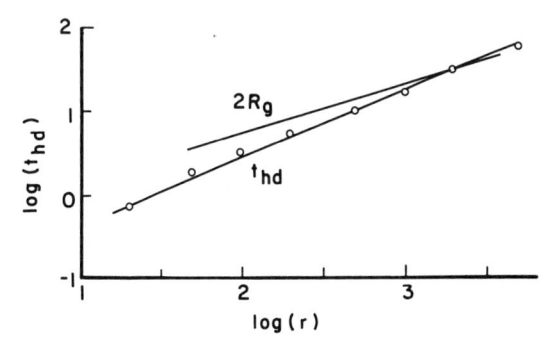

FIG. 14. Dimensionless hydrodynamic thickness vs. chain length, on a log–log scale as calculated from the SF model (Cohen Stuart et al., 1984c) for $\chi_s = 1, \chi = 0.45, \varphi_b = 10^{-4}, \lambda_0 = 0.5$, and friction per segment equivalent to a Stokes sphere of diameter $l/36$. The isolated coil diameter $2R_g = 0.17n^{0.6}$ is also shown.

polymer-coated particles, several timescales are important (de Gennes, 1982) for polymer adsorption, rearrangement of the chains on the surface, and desorption during a collision. Desorption during the short interval of a collision seems unlikely, suggesting that φ_{ads} remains constant. Scheutjens and Fleer (1985; Fleer and Scheutjens, 1982b, 1986) contrasted interaction potentials for the cases of full and restricted equilibrium.

The free energy of the two surface system is calculated from the partition function expressed in terms of Flory–Huggins solution theory. The Helmholtz free energy per site, relative to pure solvent, and at constant φ_{ads} is

$$\frac{A}{kT} = \frac{2\gamma}{kT} + \frac{\mu_p}{rkT}\sum_{i=1}^{m}\varphi_i \qquad (64)$$

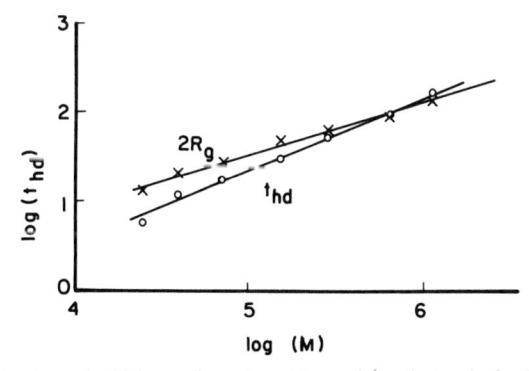

FIG. 15. Hydrodynamic thickness (in nm) vs. M, on a long–log scale, for PEO adsorbing on PS latex from water as measured by Cohen Stuart et al. (1984c). The isolated coil diameter is also shown.

with

$$\frac{2\gamma}{kT} = \sum_{i=1}^{m} \left\{ (1 - 1/r)(\varphi_i - \varphi_b) + \ln\left[\frac{1 - \varphi_i}{1 - \varphi_b}\right] + \chi[\varphi_i \langle \varphi_i \rangle - \varphi_b^2] \right\} \quad (65)$$

as the surface energy. Holding φ_{ads} constant implies that the chains' chemical potential

$$\frac{\mu_p}{kT} = 1 - \varphi_b^* - r(1 - \varphi_b^*) + \ln(\varphi_b^*) + r\chi(1 - \varphi_b^*)^2 \quad (66)$$

must vary with gap width m. The quantity

$$\varphi_b^* \equiv \frac{\sum_{i=1}^{m} \varphi_i}{\sum_{i=1}^{m} P(i, r)} \quad (67)$$

is the polymer volume fraction in a hypothetical bulk solution in equilibrium with the solution in the gap. The dimensionless interaction potential energy for restricted equilibrium,

$$V(m) = \frac{1}{kT}[A(m) - A(\infty)], \quad (68)$$

includes both the surface energy and the change in the chemical potential.

Interaction potentials for full equilibrium are always monotonically attractive. Restricted equilibrium (Fig. 16) produces more interesting behavior, with the character depending primarily on φ_{ads} and χ. Variations of χ_s and r certainly influence the results, but low concentrations of long chains or high concentrations of short chains give roughly the same adsorbed amount and interaction potential. At high φ_{ads}, the layers repel each other monotonically, suggesting "steric" stabilization; low φ_{ads} produces an attraction, perhaps due to "bridging." Figure 17 shows the maximum depth of the attraction as a function of φ_{ads} and χ.

Fleer and Scheutjens (1986) calculate the variation of the distribution of trains, loops, tails, and bridges with separation of the two surfaces. As one might expect, compression of the layers enhances the fractions of segments in trains and bridges at the expense of loops and tails. However, the existence and strength of attraction and repulsion have not been correlated with the fractions of segments in the various chain configurations.

Force measurements between polymer-coated mica surfaces exhibit many of the features predicted by the SF model. The force measurement technique characterizes the interaction between polymer layers and provides an indirect

COLLOIDAL PARTICLES AND SOLUBLE POLYMERS 177

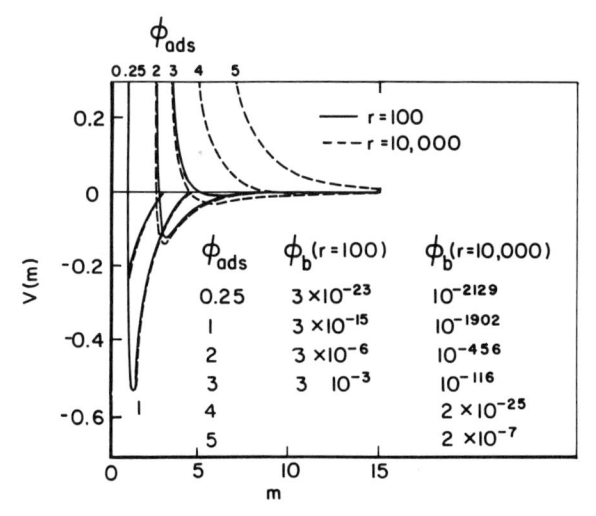

FIG. 16. Interaction potential energy between adsorbed layers vs. separation under conditions of restricted equilibrium as calculated by Scheutjens and Fleer (1985). Various adsorbed amounts are achieved for two chain lengths by varying bulk solution concentration when the surfaces are well separated; the necessary φ_b for each curve are given in the table. Other parameters include $\chi_s = 1$, $\chi = 0.5$, and $\lambda_0 = 0.5$.

measure of layer thickness. The apparatus, originally developed by Tabor and Winterton (1969), has been used extensively for investigation of van der Waals forces *in vacuo*, electrostatic forces in aqueous media, and hydration and structural forces due to ordering of solvent molecules near the surfaces. A

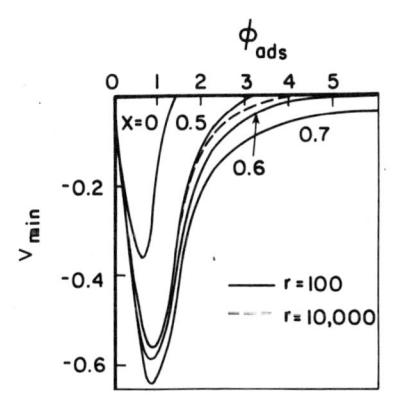

FIG. 17. Depth of the attractive minimum vs. adsorbed amount for various χ parameters and two chain lengths, as calculated by Scheutjens and Fleer (1985). No minimum implies that only repulsion is predicted.

review of these studies as well as references are provided by Russel (1987) together with references providing details of the construction and operation of the apparatus.

In all cases, the surfaces are molecularly smooth sections of cleaved mica in a crossed cylinder configuration within a thermostatted bath containing the solution of interest. The separation between the cylinders is measured with high accuracy through multiple beam interferometry. Both cylinders are attached to micrometer-driven mounts, but one of the surfaces is connected through a system of springs. Knowing the force constants of the springs, one can calculate the force between the surfaces from their separation.

In good solvents with high adsorbed amounts, PEO layers repel each other monotonically (Klein and Luckham, 1984a,b; Luckham and Klein, 1985). For strong adsorption, but low φ_{ads}, attraction is observed. To demonstrate this, Almog and Klein (1985) imposed diffusional resistances to restrict the amount of polystyrene adsorbing from cyclopentane. The attractive minima in the resultant force curves (Fig. 18) are deepest for low φ_{ads} and eventually disappear as more polymer is allowed to adsorb. Since dispersion forces are negligible in this case, only bridging can account for attraction. Similar behavior occurs with weak adsorption in slightly better-than-theta solvents (Israelachvili et al., 1984). In poor solvents (Klein, 1982, 1983), strong attraction can be ascribed to the preference of segments for other segments rather than solvent. The same features, for all solvent conditions, have been predicted by the SF model (Figs. 16 and 17) (Scheutjens and Fleer, 1985; Fleer and Scheutjens, 1986).

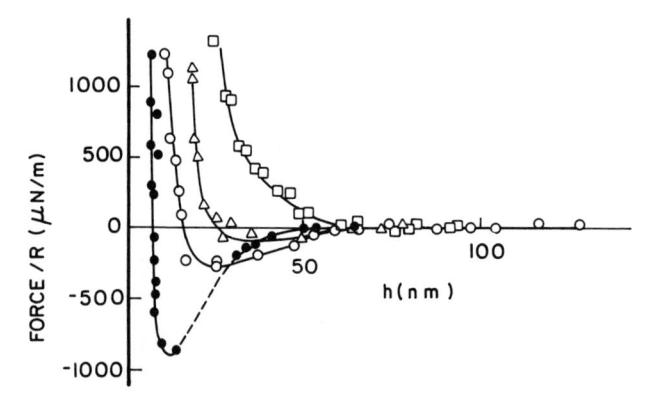

FIG. 18. Interaction potential energy (converted to a potential between planes) vs. separation between two mica surfaces bearing adsorbed polystyrene (2×10^6 g/mol) in cyclopentane, a good solvent (Almog and Klein, 1985). Full circles, open circles, triangles, and squares correspond to increasing adsorbed amounts achieved by incubating the surfaces at close separation for increasing intervals of time.

4. Conclusions

The SF model succeeds for several reasons. First, the model treats a wide range of experimental conditions without a priori assumptions about the configuration of the polymer molecules in the adsorbed layer, enumerating the statistical distribution of trains, loops, and tails to highlight the influence of particular configurations, especially tails, on experimentally observable quantities. The results indicate a relative surplus of chain ends far from the surface and depletion of ends near the surface. This conclusion agrees with earlier results of Chandrasekhar (1943) and Hesselink (1969, 1971), that adsorbed chains prefer long-tail–short-loop configurations rather than short-tail–long-loop configurations. These "end effects" significantly affect t_{hd} (Cohen Stuart et al., 1984c; Cosgrove et al., 1984) and forces between interacting adsorbed polymer layers (Scheutjens and Fleer, 1985; Fleer and Scheutjens, 1986). Extensions to polydisperse polymer and polyelectrolyte adsorption have been made and are reviewed by Fleer (1988).

On the other hand, the SF model is limited in several respects. The finite-difference form of the model facilitates numerical solution but obscures some of the underlying physics. The artificial discretization of space, necessitating the selection of λ_0 and λ_1 to characterize the lattice geometry, may not accurately reflect the reality of continuous space. Although the lattice parameters have little qualitative influence on the results (Domb, 1963), the quantitative changes significantly affect comparisons with experimental data. On a more practical level, the lattice model calculations become time consuming for long chains, especially for $n > 1000$.

F. Self-Consistent Field Theory

1. Motivation and Formulation

The SCF [Eq. (31)] forms the basis for many theories of polymers at surfaces (Dolan and Edwards, 1974, 1975; Jones and Richmond, 1977; Levine et al., 1978; Joanny et al., 1979; de Gennes, 1979) and bulk polymer interfaces (Helfand and Sapse, 1975, 1976; Helfand, 1975b,c; Helfand and Wasserman, 1976; Hong and Noolandi, 1981). In fact, the recursion relation in Eq. (56) of the SF model is simply a finite-difference form of Eq. (31); expanding $P(i + 1, s - 1)$, $P(i, s - 1)$, and $P(i - 1, s - 1)$ in i and s, rearranging, and using the identity in Eq. (57) demonstrates the equivalence. The versatility and predictive ability of the SF model suggests trying to circumvent that model's limitations and extract more physical insight by reverting to the partial differential in Eq. (31). The field equation formulation allows the use of any equation-of-state, brings to bear a variety of solution techniques, and permits extensions to important related problems.

The complete formulation includes Eqs. (31)–(33); the SCF of Eq. (39); an equation-of-state relating the local free-energy density to the volume fraction; and boundary conditions at the surface and in bulk. Unfortunately, the discussion of Section II demonstrates that no single solution theory accurately describes the thermodynamic properties for all concentrations and temperatures. Thus, specialization to particular solution models is necessary.

In addition, the local free energy density must include the energy of segmental adsorption at the surface. While lattice models simply award an additional energy χ_s to segments in the surface layer, continuum models require an assumption about the range of the potential. Ploehn et al. (1988; Ploehn and Russel, 1989) employ a "sticky surface" model (Baxter, 1969) featuring an infinitely deep and narrow attractive well which produces a finite population of segments at $z = 0$. This idealization makes $U(\mathbf{r})$, $G(\mathbf{r}, \mathbf{r}'; s)$, and $\varphi(\mathbf{r})$ discontinuous at the surface $\mathbf{r}_0 = (x, y, 0)$, but expressible as sums of continuous and discontinuous surface functions, i.e.,

$$\int G(\mathbf{r}, \mathbf{r}'; s)\,d\mathbf{r}' \equiv G_c(\mathbf{r}, s) + \delta(\mathbf{r} - \mathbf{r}_0)G_s(s)$$

$$e^{-\beta U(\mathbf{r})} \equiv e^{-\beta U_c(\mathbf{r})} + \delta(\mathbf{r} - \mathbf{r}_0)e^{-\beta U_s} \qquad (69)$$

$$\varphi(\mathbf{r}) \equiv \varphi_c(\mathbf{r}) + \delta(\mathbf{r} - \mathbf{r}_0)\varphi_s.$$

In this formulation, the starting point of a chain is integrated out for convenience.

The Flory–Huggins equation-of-state (22) provides simple expressions for the SCFs (Helfand, 1975c; Helfand and Sapse, 1975, 1976; Scheutjens and Fleer, 1979; Hong and Noolandi, 1981; Ploehn et al., 1988),

$$\beta U_c = -\ln\frac{(1 - \varphi_c)}{(1 - \varphi_b)} - 2\chi(\varphi_c - \varphi_b), \qquad z \geq 0,$$

$$\beta U_s = -\chi_s - \ln\frac{(1 - \varphi_s)}{(1 - \varphi_b)} - 2\chi(\varphi_s - \varphi_b), \qquad z = 0 \qquad (70)$$

but computer simulations suggest that the former severely underestimates the osmotic pressure of chains in a continuum (Okamoto, 1975; Croxton, 1979; Dickman and Hall, 1986), at least for athermal ($\chi = 0$) solutions. For this reason, Ploehn and Russel (1989) formulated an equation-of-state which produces greater repulsion between segments.

Supplementing the entropic repulsion of the Carnahan–Starling (1969) equation with a "van der Waals" attraction leads to the SCF (Ploehn and

Russel, 1989)

$$\beta U_c = -16\alpha(\varphi_c - \varphi_b) + \frac{\varphi_c(8 - 9\varphi_c + 3\varphi_c^2)}{(1 - \varphi_c)^3} - \frac{\varphi_b(8 - 9\varphi_b + 3\varphi_b^2)}{(1 - \varphi_b)^3} \quad (71)$$

for positions $\mathbf{r} \neq \mathbf{r}_0$. Segments and solvent constrained to the surface interact as two-dimensional entities. Augmenting the equation-of-state of Baram and Luban (1979) for hard discs with a van der Waals-type attraction and extracting the SCF via Eq. (39) gives (Ploehn and Russel, 1988)

$$\beta U_s = -\chi_s + B_0 \ln(1 - \varphi_s) + \frac{B_1 \varphi_s}{1 - \varphi_s} + \sum_{i=2}^{4} B_i(\varphi_s)^{(i-1)}$$

$$- B_0 \ln(1 - \varphi_b) - \frac{B_1 \varphi_b}{1 - \varphi_b} - \sum_{i=2}^{4} B_i(\varphi_b)^{(i-1)} \quad (72)$$

for $\mathbf{r} = \mathbf{r}_0$, where $B_0 = 3.46700$, $B_1 = 5.37944$, $B_2 = 3.67599(1 - 2\alpha) - 1.91244$, $B_3 = 0.28413$, and $B_4 = 0.016178$. Thus results for two equations-of-state can be compared to assess the influence of the details of segment interactions on the overall layer structure.

The formulation of a proper surface boundary condition is a delicate matter, as noted by DiMarzio (1965) and de Gennes (1969). Lattice models simply require that $P(i, s) \equiv 0$ for layers $i < 0$, a form proven correct by DiMarzio (1965). In continuum models, chains intersecting the surface undergo both "reflection" and "adsorption," the relative amount of each depending on the energy of contact at the surface. The result is a mixed boundary condition expressed by de Gennes (1969) as

$$\frac{1}{G} \frac{\partial G}{\partial z}\bigg|_{z=0} = -k_{\text{ads}}(T) \quad (73a)$$

for $\varphi(0) \ll 1$. For stronger adsorption, the mixed boundary condition becomes a nonlinear function of $\varphi(0)$. For example, the recursion [Eq. (56)] of the SF lattice-model applies at the surface $i = 1$ with $P(0, s) \equiv 0$. Expanding $P(2, s-1)$ and $P(1, s - 1)$ in Taylor series in i and s produces

$$\frac{P(1, s)}{P(1, 1)} = (1 - \lambda_1)\left[P - \frac{\partial P}{\partial s}\right] + \lambda_1 \frac{\partial P}{\partial i} \quad (73b)$$

as the continuous space form of their surface boundary condition. Ploehn et al. (1988) derived from Eq. (31) a general treatment of the surface region, which reduces to

$$G_c + \delta(z)G_s = e^{-\beta U}\left\{\left[\frac{1}{2} + \frac{1}{4}\mathbf{k} \cdot \nabla\right]\left(G_c - \frac{\partial G_c}{\partial s}\right) + \frac{1}{2}\left(G_s - \frac{\partial G_s}{\partial s}\right)\right\} \quad (74)$$

for $z \ll 1$. Integrating over a small interval of $O(\varepsilon)$ in z and retaining only the $O(1)$ terms, with the condition that G_c and $e^{-\beta U_c}$ vary smoothly at $z = 0$, provides the surface boundary condition

$$G_c e^{\beta U_c} = \frac{1}{2}(K_A + 1)\left[G_c - \frac{\partial G_c}{\partial s}\right] + \frac{1}{4}\mathbf{k} \cdot \nabla G_c,$$

$$G_s = K_A G_c(\mathbf{r}_0, s), \tag{75}$$

with $\mathbf{k}$ a unit vector normal to the surface and

$$K_A \equiv e^{-\beta[U_s - U_c(\mathbf{r}_0)]} \tag{76}$$

defining a partition coefficient relating segments densities at and near the surface. Since Eq. (33) requires

$$\varphi_s = K_A \varphi_c(\mathbf{r}_0), \tag{77}$$

all of the surface functions are defined in terms of their continuous counterparts. The similarity of Eqs. (75) and (73b) alludes to the relationship between adsorption energy (inherent in K_A) and chain flexibility (embodied in the lattice parameter λ_1), as demonstrated earlier by Hoeve et al. (1965).

Simple manipulations show that G_c and φ_c satisfy

$$\frac{\partial G_c}{\partial s} = \tfrac{1}{6}\nabla^2 G_c + [1 - e^{\beta U_c}]G_c \tag{78}$$

and

$$\varphi_c(\mathbf{r}) = \frac{c}{Z_A \exp(-\beta U_c)} \int_0^n G_c(\mathbf{r}, s)G_c(\mathbf{r}, n - s)\,ds \tag{79}$$

for $z > 0$; c is a proportionality constant, and Z_A is the configuration integral (defined later). Far from an adsorbing surface, chains have an equal probability of ending anywhere; hence

$$G_c(\infty, s) = 1 \tag{80}$$

merges G_c into the bulk solution. In addition, $\varphi_c(\infty) \to \varphi_b$ and $U_c \to 0$ so that Eq. (79) yields $c = \varphi_b Z_A/n$. In a sense, this procedure includes a chain configuration from bulk solution in the partition function for the adsorbed layer and implies "full" equilibrium between free and adsorbed chains.

Under some conditions, though, an adsorbed layer may not be in full equilibrium with bulk solution. For "restricted" equilibrium, the adsorbed amount of polymer is held constant; adsorbed chain configurations are assumed to be optimally distributed, but only solvent is free to move between the layer and bulk solution. Since Z_A normalizes $G(\mathbf{r}, s)G(\mathbf{r}, n - s)e^{\beta U(\mathbf{r})}$, integration of Eq. (79) with Eq. (69) over $\mathbf{r}$ gives $c = \varphi_{\text{ads}}/n$ where φ_{ads} is the constant adsorbed amount of polymer (defined later). It is convenient to define a new constant $\varphi_b^* \equiv \varphi_b$ for full equilibrium, and $\varphi_b^* \equiv \varphi_{ads}/Z_A$ for restricted equilibrium so that $c = \varphi_b^* Z_A/n$ in all cases.

2. Ground State Solutions

Various techniques enable approximate solutions of the closed system of equations; here we specialize for one dimensional variations of layer structure in the direction normal to the surface. Regarding the SCF as an external field, de Gennes (1969, 1979) treated Eq. (78) as a "pseudo-linear" equation de Gennes, 1969, 1979) and expanded G_c in eigenfunctions,

$$G_c(z, s) = \sum_{j=0}^{\infty} g_j(z) \exp(\lambda_j s) \quad (81)$$

with λ_j as the eigenvalues. For large n, the $j = 0$ (ground state) term dominates; this "ground state approximation,"

$$G_c(z, s) = g(z) \exp(\lambda_0 s), \quad (82)$$

is the basis of several analytical solutions (Jones and Richmond, 1977; Joanny et al., 1979; Ploehn et al., 1988).

Further progress requires linearizing $e^{\beta U(z)}$ as

$$e^{\beta U_c} \approx 1 + v\varphi_c + \frac{w}{2}\varphi_c^2, \quad (83)$$

with φ_b assumed to be small so that $\exp[\beta U_c(\varphi_b)] \approx 1$. Retention of the $O(\varphi^2)$ term permits application of Eq. (83) in theta solvents for which $v = 0$. The Flory–Huggins form, [Eq. (70)], yields $v = 1 - 2\chi$, and $w = 2(1 - 2\chi + 2\chi^2)$; the SCF Eqs. (71)–(72) produce $v = 8(1 - 2\alpha)$ and $w = 2[47 + 128\alpha(\alpha - 1)]$. Note that Eqs. (70)–(72) are not expanded at $z = 0$ so that φ_s and φ_c cannot exceed unity there.

Combining Eqs. (79) and (82) gives

$$\frac{\varphi_c(z)}{\varphi_b^*} = \exp(\beta U_c) g^2(z) \exp(\lambda_0 n), \quad (84)$$

and substituting Eqs. (82)–(84) into Eq. (78) produces a nonlinear ordinary differential equation for φ_c. Its solution (Ploehn et al., 1988),

$$\varphi_c(z) = \frac{4\lambda_0 c_i e^{\gamma z}}{(c_i e^{\gamma z} - v/2)^2 - \frac{2}{3}\lambda_0 w} = \frac{4\lambda_0 c_i e^{-\gamma z}}{(c_i - (v/2)e^{-\gamma z})^2 - \frac{2}{3}\lambda_0 w e^{-2\gamma z}}, \quad (85)$$

includes $\gamma = (24\lambda_0)^{1/2}$ as the reciprocal of the profile's decay length and c_i as an integration constant that must be determined from Eq. (70). A normalization condition to be discussed later fixes λ_0.

Unfortunately, the ground state approximation dictates a spatial distribution of segments along the contour of the chain, i.e.,

$$\frac{\varphi_c(z, s)}{\varphi_b^*} \equiv \frac{1}{n} G_c(z, s) G_c(z, n - s) \exp[\beta U_c(z)]$$

$$= \frac{1}{n} g^2(z) \exp(\lambda_0 n) \exp[\beta U_c(z)], \qquad (86)$$

which is *independent* of s. As in the lattice models of Roe (1974) and Levine et al. (1978), all segments are equivalent. Thus, the ground state solution in Eq. (85) ignores end effects and precludes the prediction of tails. Although the segment density profile yields some of the layer's qualitative features, the calculated values of φ_{ads}, t_{rms}, and t_{hd} fall short of the corresponding experimental quantities.

3. Matched Asymptotic Expansion

A more realistic representation of an adsorbed layer recognizes the fundamental differences between the layer structure close to and far from the surface. Such distinctions appear naturally in scaling theory through the proximal and central regions (de Gennes, 1981). In the context of mean-field theory, Scheutjens et al. (1986) demonstrated that two eigenfunctions in the solution of Eq. (78) are necessary and sufficient to describe end effects: The ground state eigenfunction dominates near the surface, while the second eigenfunction accounts for bulk solution configurations.

Ploehn and Russel (1989) developed a matched asymptotic solution of Eq. (73) equivalent to a two-eigenfunction approximation for G_c. Near the surface, $\varphi(z)$ is $O(1)$ and Eq. (81) is presumably accurate; adsorption and interchain interactions distort chain configurations so that the characteristic length is l. Far from the surface, chains are ideal and the characteristic length scales as $n^{1/2}l \gg l$. The widely separated length scales enable the "inner" ground state solution to be matched asymptotically to an "outer" solution, yielding a uniform approximation for all z.

Rescaling z and s so that $x = z/n^{1/2}$ and $t = s/n$ converts Eq. (78) to

$$\frac{\partial G_c}{\partial t} = \frac{1}{6} \frac{\partial^2 G_c}{\partial x^2} + n[1 - \exp(\beta U_c)] G_c. \qquad (87)$$

Then expanding G_c in a regular perturbation series as

$$G_c = G_0 + \frac{1}{n} G_1 + \cdots, \qquad (88)$$

and the SCF as $n(1 - e^{\beta U_c}) \simeq -nv(\varphi_c - \varphi_b)$ produces a hierarchy of independent partial differential equations, one for each power of $1/n$. Assuming that $nv(\varphi_c - \varphi_b) \ll 1$ reduces the lowest order equation to the diffusion equation,

$$\frac{\partial G_0}{\partial t} = \frac{1}{6}\frac{\partial^2 G_0}{\partial x^2}, \qquad (89)$$

expressing the ideal behavior of the outer region at this level of approximation. Homogeneity of the bulk solution requires $G_0(x, 0) = 1$ and $G_0(\infty, t) = 1$. Rescaling Eq. (75) and substituting in Eq. (88) shows that $G_0(0, t) = 0$.

Laplace transforms easily produce

$$G_0(z, s) = \text{erf}\left[\frac{z\sqrt{6}}{2\sqrt{s}}\right] \qquad (90)$$

as the outer solution; the scaling cancels since Eq. (89) is a similarity solution. The sum of G_0 and the inner (ground state) solution G_{gs} derived from Eqs. (82)–(85), minus any common terms that appear in the appropriate limits of each, then constitutes a uniformly valid approximation. However, $G_0 \to 0$ as $z \to 0$, and $G_{gs} \to 0$ as $z \to \infty$, so the common terms are zero, leaving

$$G_c(z, s) = g(z)\exp(\lambda_0 s) + \text{erf}\left[\frac{z\sqrt{6}}{2\sqrt{s}}\right] \qquad (91)$$

as the matched solution.

The integral in Eq. (79) can be evaluated analytically through Laplace transforms. Recognizing Eq. (79) as a convolution leads to

$$\mathscr{L}\left\{\frac{\varphi_c^*}{\varphi_b^*}\right\} = \tilde{G}_c^2(z, \eta),$$

with $\tilde{G}_c(z, \eta) = \mathscr{L}\{G_c(z, n)\}$. Substituting in the transform of Eq. (91) and inverting yields

$$\varphi_c(z) = \varphi_b^* \mathscr{L}^{-1}\{\tilde{G}_{gs}^2 + 2\tilde{G}_{gs}\tilde{G}_0 + \tilde{G}_0^2\}$$

with

$$\tilde{G}_{gs} = \frac{g(z)}{\eta - \lambda_0}, \qquad \tilde{G}_0 = \frac{1}{\eta}[1 - e^{-z(6\eta)^{1/2}}].$$

The corresponding volume fraction $\varphi_c(z)$ consists of three terms arising from configurations with two, one, or zero sequences of segments originating in the inner region. The first term is exactly the ground state solution, representing segments contained in loops. The second and third terms correspond to segments contained in tails and nonadsorbed chains. Labelling the individual contributions to φ_c as φ_L, φ_T, and φ_N gives

$$\varphi_c(z) = \varphi_L(z) + \varphi_T(z) + \varphi_N(z) \tag{92}$$

with

$$\varphi_L(z) = \frac{4\lambda_0 c_i e^{-\gamma z}}{\left(c_i - \frac{v}{2}e^{-\gamma z}\right)^2 - \frac{2}{3}\lambda_0 w e^{-2\gamma z}}$$

$$\frac{\varphi_T(z)}{\varphi_b^*} = \frac{2g(z)}{n\lambda_0}\left\{e^{n\lambda_0}\left[1 - \frac{1}{2}(e^{-(6\lambda_0)^{1/2}z})\right]\text{erfc}\left[\frac{z\sqrt{6}}{2\sqrt{n}} - (n\lambda_0)^{1/2}\right]\right. \tag{93}$$

$$\left. + e^{(6\lambda_0)^{1/2}z}\text{erfc}\left[\frac{z\sqrt{6}}{2\sqrt{n}} + (n\lambda_0)^{1/2}\right] - \text{erf}\left[\frac{z\sqrt{6}}{2\sqrt{n}}\right]\right\},$$

[$G(z)$ from Eq. (84) with φ_c replaced by φ_L and $\exp(\beta U_c) \approx 1$] and

$$\frac{\varphi_N}{\varphi_b^*} = 1 - 2\left[1 + \frac{3z^2}{n}\right]\text{erfc}\left[\frac{z\sqrt{6}}{2\sqrt{n}}\right] + 2\left[\frac{6}{\pi n}\right]^{1/2} z e^{-3z^2/2n}$$

$$+ \left[1 + \frac{12z^2}{n}\right]\text{erfc}\left[\frac{z\sqrt{6}}{\sqrt{n}}\right] - 2\left[\frac{6}{\pi n}\right]^{1/2} z e^{-6z^2/n}.$$

Finally, the solution must be normalized in order to determine λ_0. The adsorbed amount (on a per surface basis),

$$\varphi_{\text{ads}} \equiv \int_0^\infty [\varphi_c(z) - \varphi_N(z)][1 + \delta(z)K_A] dz$$

$$\equiv \frac{\varphi_b^*}{n}\int_0^\infty \int_0^n [G_{\text{gs}}(z, s)G_{\text{gs}}(z, n - s) + G_{\text{gs}}(z, s)G_0(z, n - s)$$

$$+ G_0(z, s)G_{\text{gs}}(z, n - s)][1 + \delta(z)K_A] ds dz, \tag{94}$$

includes only chains having at least one segment on the surface. Alternately, a count of chain ends multiplied by n gives φ_{ads} since every chain has an end:

$$\varphi_{\text{ads}} = \varphi_b^* \int_0^\infty [G(z, n) - G_0(z, n)] dz \tag{95a}$$

gives this tally. For full equilibrium, λ_0 and φ_{ads} must satisfy Eqs. (94) and (95a).

The configuration integral for adsorbed chains is defined as

$$Z_A \equiv \int_0^\infty [G(z, n) - G_0(z, n)] dz = \int_0^\infty G_{\text{gs}}(z, n)[1 + \delta(z)K_A) dz \tag{95b}$$

so that nonadsorbed chains are not included. Under restricted equilibrium conditions, $\varphi_b^* = \varphi_{ads} n/Z_A$ and Eq. (95a) is an identity; however, Eq. (94) can be taken as another equation for Z_A. Thus Eqs. (94) and (95b) are solved for λ_0 and Z_A when φ_{ads} is held constant.

4. Results for Isolated Layers

Apart from two constants, c_i and λ_0, found as the zeroes of two nonlinear algebraic equations, the volume fraction profiles expressed in Eq. (93) are analytical and easily evaluated. Typical profiles shown in Fig. 19 (Ploehn and Russel, 1989) roughly correspond to 5.3×10^3 kg/mol polystyrene adsorbing from cyclohexane at $T = \theta$ with $\chi_s = 1.0$. Scaling φ on φ_b and plotting on a logarithmic scale magnifies the details at low φ relative to the linear scale in the inset. Distance from the surface is scaled with the radius of gyration of an ideal chain, $R_g = (nl^2/6)^{1/2}$.

The volume fraction falls off sharply over $0 \le z \le \frac{1}{3} R_g$ and is dominated by the exponential decay of segments contained in loops. In the interval $0.80 R_g < z < 1.80 R_g$, φ decays more slowly with tail segments making up

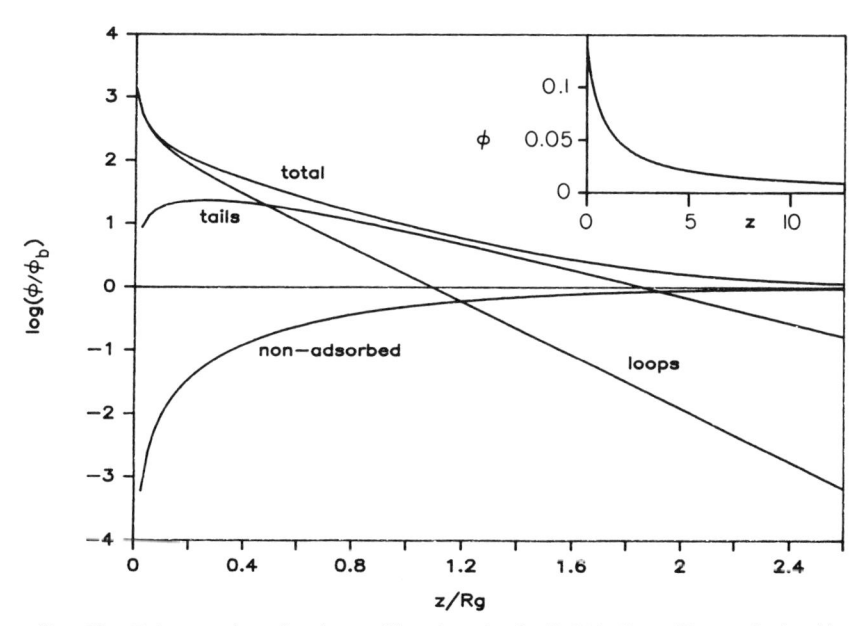

FIG. 19. Polymer volume fraction profiles, plotted as $\log[\varphi(z)/\varphi_b]$ vs. z/R_g, as calculated by Ploehn and Russel (1989) using the SCF given by Eqs. (71) and (72). The profiles for all segments (total), loops, tails, and nonadsorbed segments correspond to φ, φ_L, φ_T, and φ_N given by Eqs. (92) and (93). The total profile plotted on a linear scale is shown in the inset.

most of the total. As found by Scheutjens and Fleer (1980), the tail profile has a maximum and then falls off exponentially, but with a much longer decay length than the loop profile. Beyond about 2.5 R_g, nonadsorbed segments comprise most of total as the adsorbed layer merges into the bulk solution. Segments near the surface almost always belong to adsorbed chains, while segments of nonadsorbed chains are depleted there. In a sense, the surface is a "sink" for nonadsorbed chain probability and a "source" of adsorbed chain probability.

The features of $\varphi(z)$ obtained from the matched asymptotic expansion closely resemble those predicted by the SF model (Fig. 11) and measured through neutron scattering (Cosgrove et al., 1987a). Clearly, mean-field models can account for the effects of adsorption energy, solvency, and excluded volume on layer structure. Effects arising from complex phenomena such as surface heterogeneity or molecular details such as polar association among segments cannot be accommodated. Such complications are suspected, for example, for PEO solutions and layers (Israelachvili et al., 1980) as probed with neutron scattering by Cosgrove et al. (1987b); hence, behavior beyond the scope of the theories might be responsible for some of the quantitative disagreement between theory and experiment.

The adsorbed amount, defined by Eq. (94), is related to the experimental polymer adsorbance Γ (g/cm^2) through $\varphi_{ads} = \Gamma \tilde{v}/l$. Adsorption isotherms (Ploehn et al., 1988; Ploehn and Russel, 1989) show a sharp initial rise of φ_{ads} followed by a plateau as φ_b increases. The plateau value of φ_{ads} is less in a good solvent than in a theta solvent. These trends are widely observed experimentally and theoretically (Takahashi and Kawaguchi, 1982; Cohen Stuart et al., 1986; Fleer and Lyklema, 1983).

The predicted influence of n on φ_{ads} is compared in Fig. 20 with experimental data for adsorption of polystyrene on chrome (Takahashi et al., 1980) and silica (vander Linden and van Leemput, 1978; Kawaguchi et al., 1980) from cyclohexane at $T = \theta$. Curves A and B ($\chi_s = 1$ and 15) were calculated using Eqs. (71) and (72); curves C and D ($\chi_s = 1$ and 15) are based on the Flory–Huggins SCF of Eq. (70). Except for χ_s, all parameters are calculated from experimental conditions.

These results indicate that φ_{ads} increases with χ_s and n but appears to reach a limit at large n in good solvents. φ_{ads} is roughly linear in log(n) for $n < 10^4$. Selection of $\chi_s = 1.0$ produces curves (A and C) having the same qualitative features as the silica data. Although the predictions with two equations-of-state bracket the silica data, neither produces quantitative agreement over a significant range of n. Increasing χ_s to 15 (curves B and D) produces limited agreement with *different* sets of data, but loses some of the qualitative features of the data for $n < 1000$. Values of $\chi_s > 15$ are probably unrealistic.

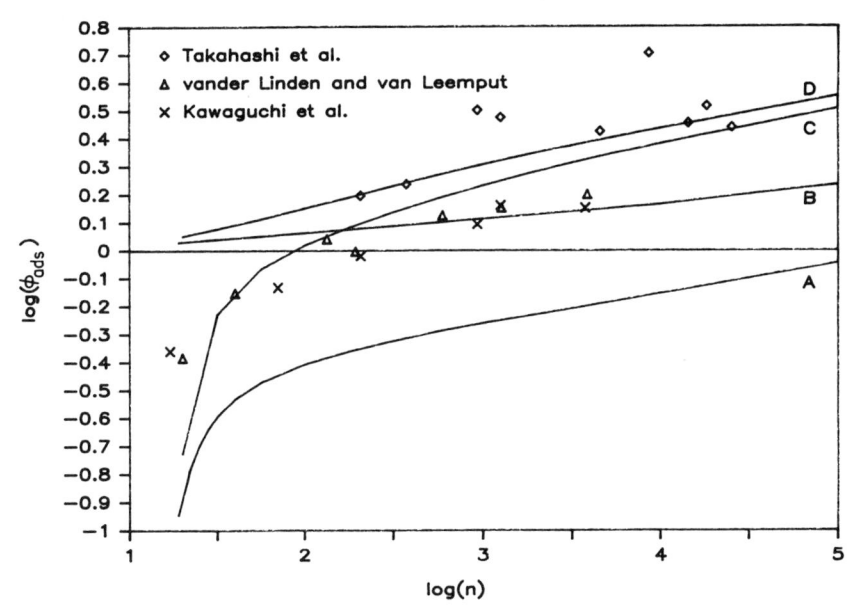

FIG. 20. Adsorbed amount vs. chain length plotted on a log–log scale. The points are measured values for PS/chrome/cyclohexane (diamonds, Takahashi et al., 1980) and PS/silica/cyclohexane (triangles, vander Linden and van Leemput, 1978; crosses, Kawaguchi et al., 1980). The curves are the predictions of Ploehn and Russel (1988) for $\chi = 0.5$ and $\varphi_b = 1.856 \times 10^{-4}$. Curves A and B ($\chi_s = 1$ and 15) are based on the SCF of Eqs. (71) and (72); curves C and D ($\chi_s = 1$ and 15) utilize the Flory–Huggins SCF of Eq. (70).

Ellipsometry measures the relative attenuation and phase shift of polarized light reflected from a polymer-coated surface. The Drude equations (Drude, 1889a,b, 1890; Stromberg et al., 1963; McCrackin and Colson, 1964) relate the attenuation and phase shift to the average refractive index and thickness t_{el} of an equivalent homogeneous film. Interpretation of t_{el} in terms of the actual refractive index distribution or the polymer distribution $\varphi(z)$ is problematic. The ad hoc analysis of McCrackin and Colson (1964) provides

$$t_{el} = \frac{\varphi_{ads}^2}{\int_0^\infty [\varphi(z) - \varphi_h]^2 \, dz}, \qquad (96)$$

and other calculations show that t_{el} is proportional to t_{rms} if various profiles for $\varphi(z)$ (e.g., Gaussian, exponential, linear) are assumed. Rigorous solutions of Maxwell's equations by Charmet and de Gennes (1983), accurate to first order in the refractive index difference between polymer and solvent, yield

$$t_{el} = \frac{2 \int_0^\infty z[\varphi_c(z) - \varphi_N(z)] \, dz}{\varphi_{ads}}, \qquad (97)$$

which gives the proper thickness for a homogeneous film of finite extent.

Extensive experiments (Takahashi and Kawaguchi, 1982) show $t_{el} \approx M^a$, where $a = 0.50$ for theta solvents and $a = 0.40$ for good solvents. Figure 21 illustrates the power-law scaling for polystyrene adsorption onto chrome from cyclohexane at $T = \theta \approx 35°C$ (Takahashi et al., 1980) and from CCl_4 at $T = 35°C$, a good solvent (Kawaguchi et al., 1983). A least-squares fit of the two sets of data gives slopes of 0.46 and 0.44 for the theta and good solvents, respectively. Figure 21 also displays several different theoretical predictions for the variation of $\log(t_{el})$ with $\log(n)$ (Ploehn, 1988); all are based on Eq. (97). Curves A and B ($\chi = 0.4993$ and 0.452) derive from the matched asymptotic solution with the SCF in Eqs. (71)–(72). For the theta solvent

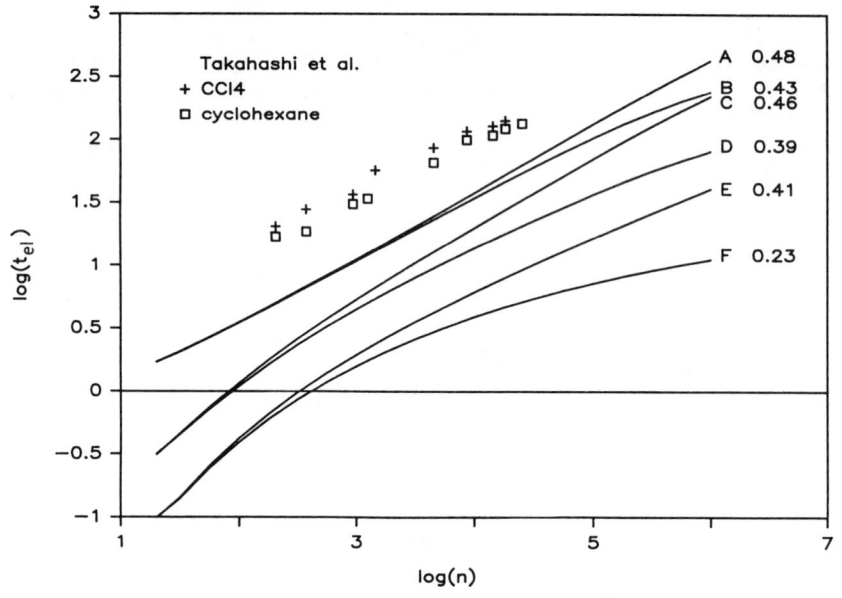

FIG. 21. Ellipsometric thickness as a function of chain length plotted on a log–log scale. The points (squares and crosses) are the data of Takahashi et al. (1980) for PS adsorbing onto chrome from cyclohexane or CCl_4. Curves A and B are calculated using the SCF in Eqs. (71) and (72); curves C–F utilize the SCF of Eq. (70). Curves A–D result from the matched asymptotic solution, while curves E and F are groundstate solutions. Other parameters include $\chi_s = 1$ and $\varphi_b = 2.784 \times 10^{-3}$. Numbers on the right are estimated slopes.

case (curve A), the calculated values are within a factor of 2–3 of the measurements, and the slope, 0.48, agrees, within error limits, with the experimental value. However, the predictions for the good solvent case (curve B) lie below those for the theta solvent, the opposite of the experimental trend. The calculated slope, 0.43, does match the experimental value. These predictions represent a considerable improvement over those based on the Flory–Huggins SCF in Eq. (70): Curves C and D ($\chi = 0.4993$ and 0.452) from the matched asymptotic solution and curves E and F ($\chi = 0.4993$ and 0.452) from on the ground state solution.

Figure 21 suggests some conclusions regarding the effect of individual contributions in the matched asymptotic solution and the SCF [Eqs. (71)–(72)] relative to the ground-state/Flory–Huggins field solution. First, the improvement of curves C and D over curves E and F stems from the matched asymptotic solution alone, i.e., the inclusion of tails, since the field is the same. The exponents should be regarded with some skepticism, though, since some of the lines are curved. Changing the field from Eq. (70) to Eqs. (71)–(72) produces power-law scaling, better exponents, and closer quantitative agreement with the data. Finally, the thicknesses for good and theta solvents move together, at least for low n.

The calculated hydrodynamic thickness, t_{hd}, shows similar agreement (Fig. 22) with the thicknesses measured by Cohen Stuart et al. (1984c) and Kato et al. (1981) by dynamic light scattering for adsorption of PEO on polystyrene particles suspended in water at 25°C (a very good solvent for PEO). Both groups find $t_{hd} \approx n^a$ with $a = 0.80$ (Cohen Stuart et al., 1984c) or $a = 0.56$ (Kato et al., 1981). The values of t_{hd} calculated by Ploehn and Russel (1989) show the same scaling ($a = 0.66$) but only half the magnitude. The frictional drag of each segment equals that of a sphere of diameter l or $2l$; the results vary relatively little with the segmental drag or χ_s.

Figure 22 also demonstrates that the hydrodynamic thickness due to segments in loops alone is considerably less than that due to all segments. Indeed, previous calculations (Cohen Stuart et al., 1984c; Anderson and Kim, 1987; Scheutjens et al., 1986) indicate that hydrodynamic methods probe the outer region of the adsorbed layer where tails dominate.

5. Interactions Between Layers

The SCF [Eq. (39)] minimizes the free energy of the interfacial system. The Helmholtz free energy per area of segment for restricted equilibrium is

$$\frac{A}{kT} = \frac{2\gamma}{kT} + \frac{\mu_p \varphi_{ads}}{nkT}, \qquad (98)$$

with the surface energy (equivalent to the interaction potential for full

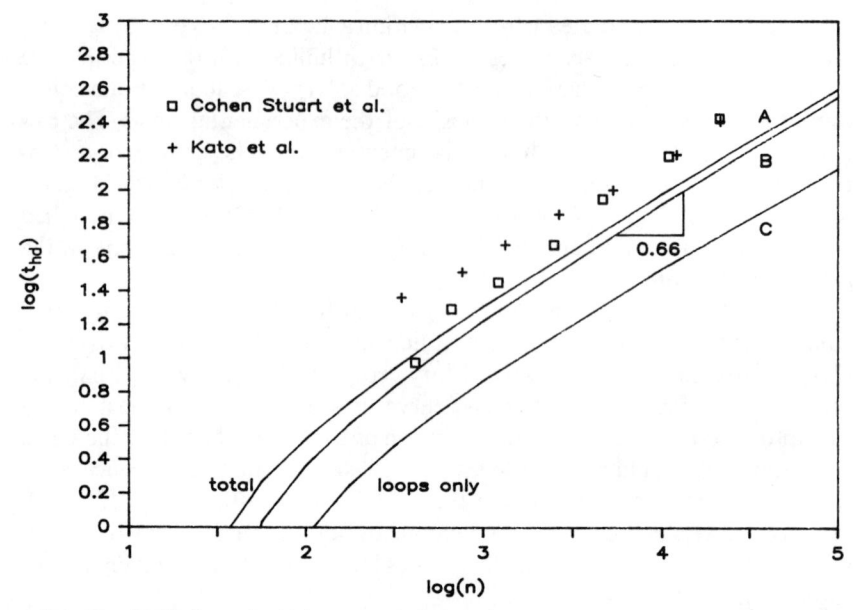

FIG. 22. Hydrodynamic thickness vs. chain length plotted on a log–log scale. The points (squares and crosses) are data for PEO/PS latex/water measured through dynamic light-scattering (squares, Cohen Stuart et al., 1984c; crosses, Kato et al., 1981). The curves are calculated by Ploehn and Russel (1989) using the SCF of Eqs. (71) and (72). Curves A and B denote frictions per segment equivalent to Stokes spheres of diameter l and $2l$, respectively. Curve C is the thickness based only on segments contained in loops (Stokes sphere diameter = l).

equilibrium) expressed as

$$2\gamma = -l^2 \int_0^{z_m} [\Pi_m(z) - \Pi_b]\, dz, \qquad (99)$$

where z_m is the dimensionless gap half-width (infinite for a single surface), $\Pi_m(z) \equiv \Pi[\varphi(z; z_m)]$ is the osmotic pressure of a solution with volume fraction φ, and Π_b is that of bulk solution (Helfand, 1975a; Hong and Noolandi, 1981; de Gennes, 1981; Ploehn, 1988). Equation (98) is equivalent to Eq. (64) as used by Scheutjens and Fleer (1985). Since the polymer distribution is discontinuous at the surface, the osmotic pressure difference includes continuous and surface functions, e.g.,

$$\Pi_m - \Pi_b = \Pi_c - \Pi_b + \delta(z)[\Pi_s - \Pi_b], \qquad (100)$$

with Π_c and Π_s provided by a polymer solution equation-of-state.

The Flory–Huggins forms of γ and μ_p are simply the continuous space analogs of Eqs. (65) and (66). Ploehn and Russel (1989) use a segment equation-of-state giving greater intersegment repulsion than the Flory–Huggins equation. Applying the prescription of Dickman and Hall (1986) produces the *polymer* equation-of-state

$$\frac{\Pi_c v_1}{kT} = \lambda_n \left[\frac{\varphi_c(1 + \varphi_c + \varphi_c^2 + \varphi_c^3)}{(1 - \varphi_c)^3} - 8\alpha \varphi_c^2 \right] + \varphi_c/n(1 - n\lambda_n)$$

$$\frac{\Pi_s v_1}{kT} = \lambda_n \left[B_0 \ln(1 - \varphi_s) + \frac{B_1 \varphi_s}{1 - \varphi_s} + (B_1 - B_0 + 1)\varphi_s + \tfrac{1}{2} B_2 \varphi_s^2 \right.$$

$$\left. + \tfrac{2}{3} B_3 \varphi_s^3 + \tfrac{3}{4} B_4 \varphi_s^4 \right] + \frac{\varphi_s}{n}(1 - n\lambda_n), \tag{101}$$

where $v_1 \equiv \tilde{v} M/n N_A$ is the volume per segment, $\lambda_n = 0.61418 + 0.45914/n$, and the B_i are defined following Eq. (72). With

$$\varphi_b^* \equiv \frac{\varphi_{ads}}{Z_A} = \frac{\varphi_{ads}}{\int_0^{z_m} [G(z, n) - G_0(z, n)]\, dz}$$

for restricted equilibrium (analogous to Eq. (67)], the chemical potential of chains confined to the gap is

$$\frac{\mu_p}{kT} = 1 + \ln(\varphi_b^*) + n\lambda_n \left[\frac{3 - \varphi_b^*}{(1 - \varphi_b^*)^3} - 16\alpha \varphi_b^* \right]. \tag{102}$$

Interaction potential energies for restricted equilibrium based on Ploehn and Russel's (1989) equations are given in dimensionless form as

$$V(z_m) = \frac{v_1}{l^3 kT} [A(z_m) - A(\infty)]. \tag{103}$$

Like the models of de Gennes (1982) and Scheutjens and Fleer (1985; Fleer and Scheutjens, 1986), the SCF model predicts monotonic attraction between adsorbed layers under conditions of full equilibrium. For constant φ_{ads} (restricted equilibrium), Fig. 23 shows $\Delta \gamma \equiv \gamma(z_m) - \gamma(\infty)$ (curve A) increases slightly before falling as the separation decreases; $\Delta \mu_p \equiv \mu_p(z_m) - \mu_p(\infty)$ increases upon compression (curve B), and the total potential (curve C) displays an attractive minimum as well as a steep repulsive "wall." Potentials for various φ_{ads} (Fig. 24) resemble those of Fig. 16 and those measured by Almog and Klein (1985) (Fig. 18). In contrast to the SF model, the potentials do not superimpose for the same φ_{ads} with different combinations of n and φ_b (i.e., dosage at infinite separation). The reasons for this difference are not clear at this point.

The parameters for the predictions presented in Fig. 24 are intended to duplicate the conditions used by Almog and Klein (1985) for adsorption of 6×10^5 g/mol polystyrene (the data of Fig. 18 are for 2×10^6 g/mol).

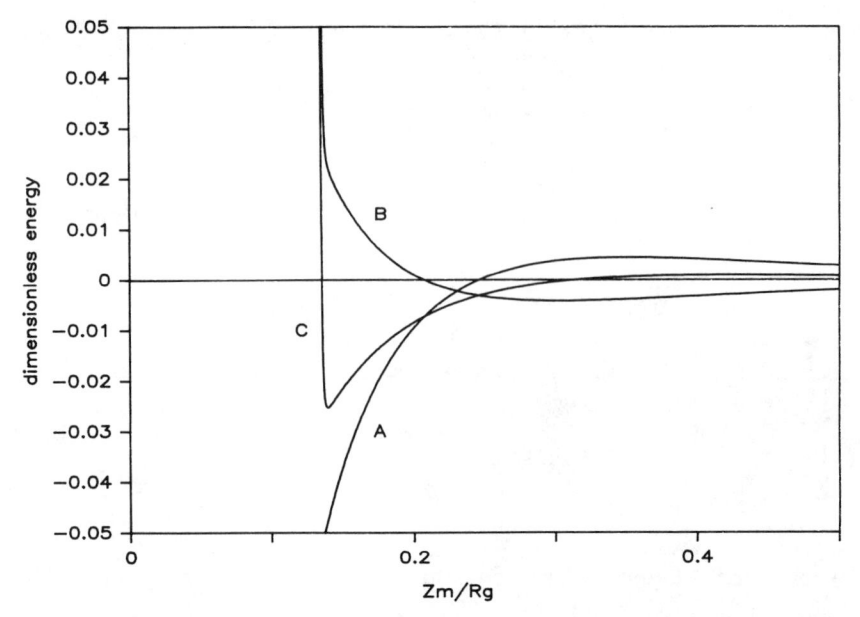

FIG. 23. Typical interaction potential energy vs. gap half-width (scaled on R_g) for restricted equilibrium (curve C) calculated by Ploehn (1988) using the Flory–Huggins SCF of Eq. (70). The components of the potential are the potential for full equilibrium, i.e., the surface tension, (curve A) and the chains' chemical potential (curve B). Parameters are $\chi_s = 1$, $\chi = 0.488$, and $n = 1129$; the dosage $\varphi_b = 1.392 \times 10^{-10}$ at large separation gives $\varphi_{ads} = 1.308$, which is held constant.

After incubating the mica surfaces at a large separation in 15 μg/ml solution, Almog and Klein reported $\Gamma = 2.5 \pm 1.5$ mg/m², equivalent to $\varphi_{ads} = \tilde{v}\Gamma/l = 1.58 \pm 0.95$ (Ploehn, 1988) with $\tilde{v}$ as the specific volume of the polymer given in Table I. Using $\varphi_b(\infty) = c_b\tilde{v} = 1.392 \times 10^{-5}$ (Ploehn, 1988), the matched asymptotic solution for a single surface predicts $\varphi_{ads} = 1.583$ agreeing with the measured value within experimental error. For this case, Almog and Klein measured a purely repulsive potential, but the corresponding theoretical prediction, curve B of Fig. 24, displays some attraction. However, the measured location of the repulsive wall at $2z_m l = 30$ nm, or $z_m l / R_g = (30 \text{ nm}/2)(38.5 \text{ nm}) = 0.39$ compares favorably with the predicted location.

Almog and Klein measured attractive minima as deep as 1000 μN/m; multiplying by $v_1/2\pi l kT$ (the 2π arises in converting the potential between the curved mica surfaces to that between planes) gives dimensionless potentials as large as 0.1. The same order of magnitude results from the SF model (Figs. 16 and 17) when v_1/l is taken as 1 nm². The matched asymptotic solution using the Flory–Huggins equation-of-state (Fig. 24) yields attractive minima of $O(0.01)$; using $v_1/l \approx 0.5$ nm² (Ploehn, 1988), the attractions also agree with the

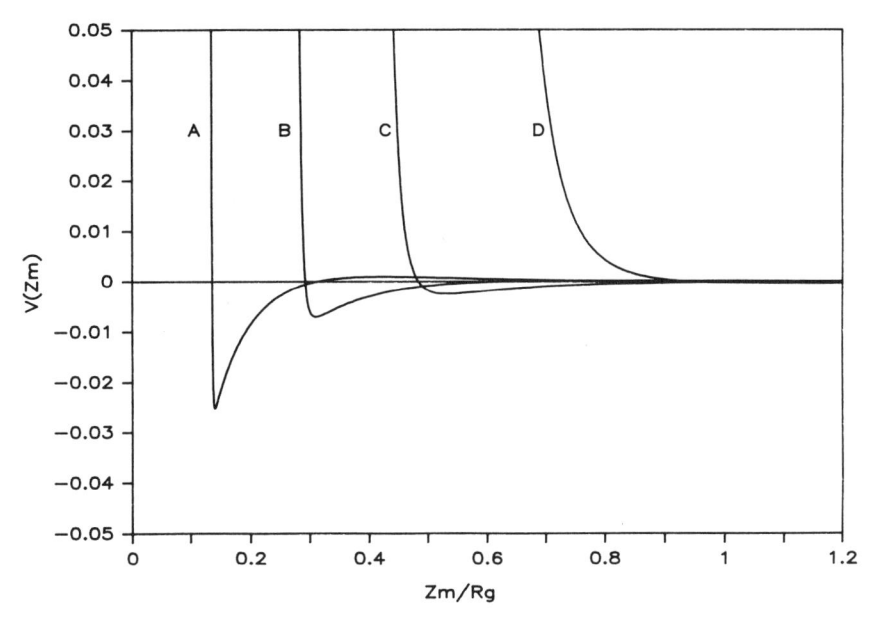

FIG. 24. Interaction potential energy vs. gap half-width (scaled on R_g) for various φ_{ads} as calculated by Ploehn (1988) using the SCF of Eq. (70). Parameters include $\chi_s = 1$, $\chi = 0.488$, and $n = 1129$. Curves A–D denote increasing dosages at large separations: $\varphi_b = 1.392 \times 10^{-10}$, 3.65×10^{-5}, 1.392×10^{-3}, and 1.392×10^{-2} correspond to $\varphi_{ads} = 1.308$, 1.583, 1.825, and 2.232, respectively.

magnitude of the measured values. However, the equation-of-state, Eqs. (71) and (72), which produces Eqs. (101) and (102) leads to potentials with much weaker attractive minima (Fig. 25) under the same conditions. The reasons for the differences between the results in Figs. 24 and 25 are entirely due to the different SCFs. The reasons for the differences between the theory and experimental data are not yet clear but are currently being investigated. The work of Evans (1989) suggests that the potential of Eq. (103) may not be appropriate for comparisons with the experimental force measurements. Rather, a work potential based on an integral of the total *stress* from infinite separation to z_m is indicated. Evans' calculations for interacting depletion layers at full equilibrium must be extended to the adsorption case with restricted equilibrium in order to resolve this important point.

One should realize that these forces, measured between macroscopic surfaces, often exceed those relevant for interacting colloidal particles. Even a small attraction between planes, say, of $O(10^{-3}\ kT/\text{nm}^2)$, translates into an attraction of $O(10kT)$ between 0.1 μm spheres separated by a distance of $O(1R_g)$. Thus the lower end of the experimental range is of primary interest.

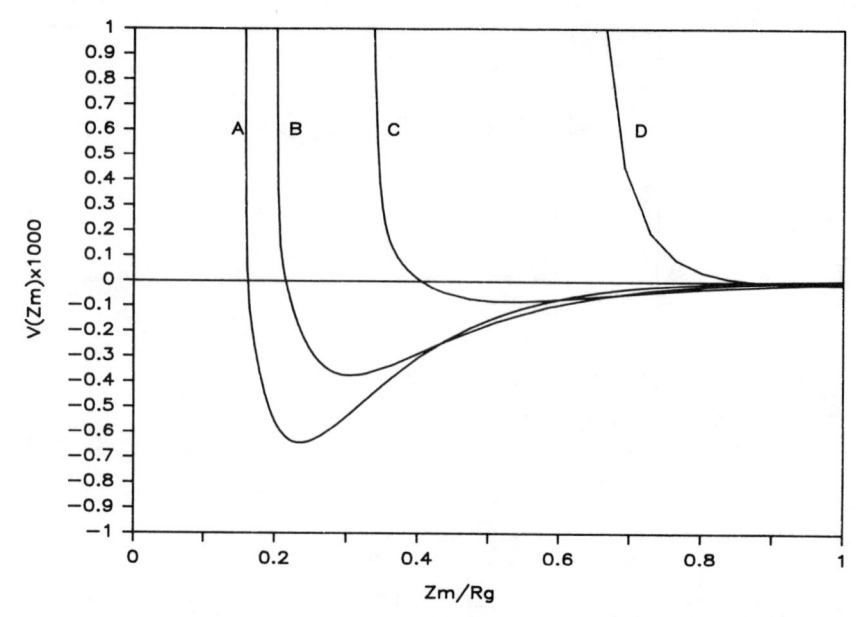

FIG. 25. Interaction potential energy vs. gap half-width (scaled on R_g) for various φ_{ads} as calculated by Ploehn (1988) using the SCF of Eqs. (71) and (72). Parameters include $\chi_s = 1$, $\chi = 0.488$, and $n = 1129$. Curves A–D denote increasing dosages at large separations: $\varphi_b = 1.000 \times 10^{-10}$, 1.392×10^{-8}, 1.392×10^{-5}, and 1.392×10^{-3} correspond to $\varphi_{ads} = 0.449$, 0.468, 0.515, and 0.613, respectively.

6. Conclusions

A matched asymptotic solution of the general SCF equations, a mathematical approximation of the exact numerical solution furnished by the SF model, provides analytical solutions and successfully predicts many of the qualitative features of experimental data. The SF model suffers the approximations inherent in Flory–Huggins statistics and the limitations imposed by the iterative numerical solutions, but does encompass all eigenfunctions rather than just two. The two-eigenfunction solution of Ploehn et al. underestimates the measured φ_{ads} and t_{el} by a considerable margin; quantitative comparisons have not yet been made with the Schuetjens–Fleer (1980) predictions. The clearest success of the Ploehn–Russel (1988) model is the prediction of t_{hd}; two eigenfunctions encompass tails that apparently dominate the hydrodynamics of adsorbed layers. Interaction potentials between adsorbed layers show the same characteristics as those measured between polymer-coated mica surfaces, although not in quantitative agreement.

Adsorption models are only as accurate as the local free-energy density that determines the SCF. A limited number of comparisons (Figs. 18, 24, and 25) between results based on the Flory–Huggins SCF, and those generated from

Ploehn and Russel's (1988) SCF, clearly show different characteristics for the adsorbed layer. Further progress depends on the evolution of accurate polymer solution theories as well as more extensive comparisons of solution and adsorption data with theoretical predictions.

G. Summary

Extensive experimentation demonstrates that polymer adsorption differs fundamentally from small molecule adsorption because of the polymer's configurational degrees of freedom. Increasingly sophisticated theories have emerged to predict the characteristics of adsorbed layers and their interactions. Early theories assumed isolated chains fixed at the surface, but recent efforts incorporate polymer excluded-volume, finite molecular weight, and equilibrium with bulk solution. Thus, quantities of practical interest, such as adsorbed amount, bound fraction, layer thicknesses, and interaction potentials, can be predicted qualitatively as functions of molecular weight, bulk solution concentration, and temperature. Improvements in quantitative comparisons between theory and experiment rely on (1) the development of new model colloidal materials and polymers with controlled microstructure and chemistry, (2) advances in experimental techniques for probing adsorbed layers, and (3) more realistic models of polymer solutions and polymer–surface interactions. Accurate prediction of large scale properties seems within reach.

IV. Terminally Anchored Polymers

A. Isolated Layers

When soluble polymers are attached by one end to a surface, the thickness of the resulting layer, L, depends on the surface density of chains σ as well as n and the excluded volume v/l^3 (de Gennes, 1980). At low densities $\sigma \langle r^2 \rangle < 1$, the isolated chains extend $\sim \langle r^2 \rangle^{1/2}$ into the solution, creating a layer with the density profile shown in Fig. 26a and a thickness of $L = n^{1/2} l$ for ideal chains and $L \sim n^{3/5} l$ in good solvents. When $\sigma \langle r^2 \rangle > 1$ the coils overlap and the interactions will cause the chains to expand away from the surface into the bulk. The configurations of the individual molecules and the density profile within the layer (Fig. 26b) differ markedly from the dilute situation. When $\sigma l^2 \sim 1$ the molecules become fully stretched.

For $\sigma \langle r^2 \rangle > 1$, the polymer chains interpenetrate, and segment densities become independent of lateral position and amenable to treatment by the SCF theory (Dolan and Edwards, 1974, 1975). The number of configurations of a

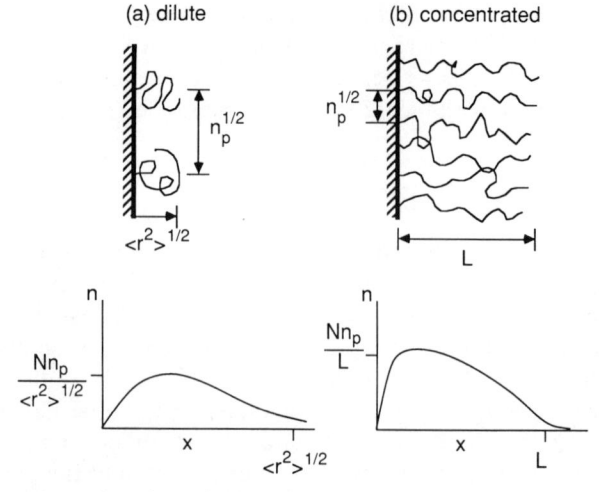

FIG. 26. Conformation of terminally anchored chains: (a) Isolated coils at low coverages (de Gennes, 1980); (b) high stretched chains at high coverages (Milner et al., 1988). The distance between graph points scales as $\sigma^{-1/2}$.

chain of length s beginning at z' and ending at z, $G(z, z', s)$, is governed by Eq. (31) with $G(z, z', 0) = \delta(z - z')$ and

$$\rho(z) = \sigma \frac{\int_0^n G(z, 0, s) \int_0^\infty G(z, z'', n - s) \, dz'' \, ds}{\int_0^\infty G(z'', 0, n) \, dz''}. \tag{104}$$

The boundary condition,

$$\frac{\partial G}{\partial s} = \frac{l}{2} \frac{\partial G}{\partial z} - \left[2 \exp\left(\frac{U}{kT}\right) - 1 \right] G \tag{105}$$

at $z = 0$, follows from Eq. (74). However, $\partial \ln G / \partial s \ll 1$ and $U/kT \ll 1$ near the interface, reducing this to

$$0 = \frac{l}{2} \frac{\partial G}{\partial z} - G, \tag{106}$$

which is equivalent to setting $G(-l/2) = 0$ or letting $G(0) = 0$ and considering the chains to be attached at $z' = l/2$.

For ideal conditions, i.e., $U = 0$, the solution to Eq. (31) (Dolan and Edwards, 1974),

$$G(z, z', s) = \left(\frac{3}{2\pi l^2 s}\right)^{1/2} \left\{ \exp\left(-\frac{3(z - z')^2}{2l^2 s}\right) - \exp\left(-\frac{3(z + z')^2}{2l^2 s}\right) \right\}, \tag{107}$$

corresponds to individual Gaussian chains anchored to the surface and determines the mean square end-to-end distance and free energy of each chain as

$$\langle z^2 \rangle = \tfrac{2}{3} nl^2$$
$$\frac{A}{kT} = \frac{1}{2} \ln \frac{2\pi n}{3}. \tag{108}$$

Since $\langle z^2 \rangle = \langle r^2 \rangle / 3 = nl^2/3$ and $A/kT = 0$ for ideal chains in solution, the presence of the wall increases both $\langle z^2 \rangle$ and A/kT.

The effect of solvent quality on the thickness of the layer L and free energy of the individual chains is demonstrated by the work of Milner et al. (1988) and Halperin (1988). The result for the chain configurations comprises an asymptotic solution of Eq. (31) for highly stretched chains, i.e., $L \gg n^{1/2} l$, via the WKB approximation. This yields the end segment probability as

$$\frac{1}{W} G(z, 0, n) = \frac{\pi^2}{4} \frac{l^3}{n^3 v} z (L^2 - z^2)^{1/2} \tag{109}$$

for $0 \leq z \leq L$ and a SCF which varies quadratically with distance from the surface as

$$\frac{U}{kT} = v\rho + \tfrac{1}{2} w\rho^2$$
$$= \frac{\pi^2}{8(nl)^2} (L^2 - z^2). \tag{110}$$

Solving for $\rho(z)$ and then integrating the profile with the condition that

$$\int_0^L \rho(z)\, dz = n\sigma \tag{111}$$

determines the dimensionless layer thickness $\bar{L}$ through

$$\bar{v}^{-2} = \pm (1 + \bar{L}^2) \sin^{-1}\left(\frac{\bar{L}}{(1 + \bar{L}^2)^{1/2}}\right) - \bar{L} \tag{112}$$

with $\bar{L} = \pi w^{1/2} L / 2nvl$ and $\bar{v} = v(l/\pi w^{3/2} \sigma)^{1/2}$.

The numerical results for $\bar{L}$ in Fig. 27 illustrate the effect of solvent quality on the layer thickness. The limiting forms,

$$\bar{v} \gg 0, \quad L \sim \left(\frac{12v\sigma}{\pi^2 l}\right)^{1/3} nl$$

$$\bar{v} \sim 0, \quad L \sim \frac{2}{\pi} \left(\frac{2w^{1/2}\sigma}{l}\right)^{1/2} nl + \left(\frac{2}{\pi^2}\right)\left(\frac{v}{w^{1/2}}\right) nl \tag{113}$$

$$\bar{v} \ll 0, \quad L \sim -\frac{w\sigma}{vl}(nl)$$

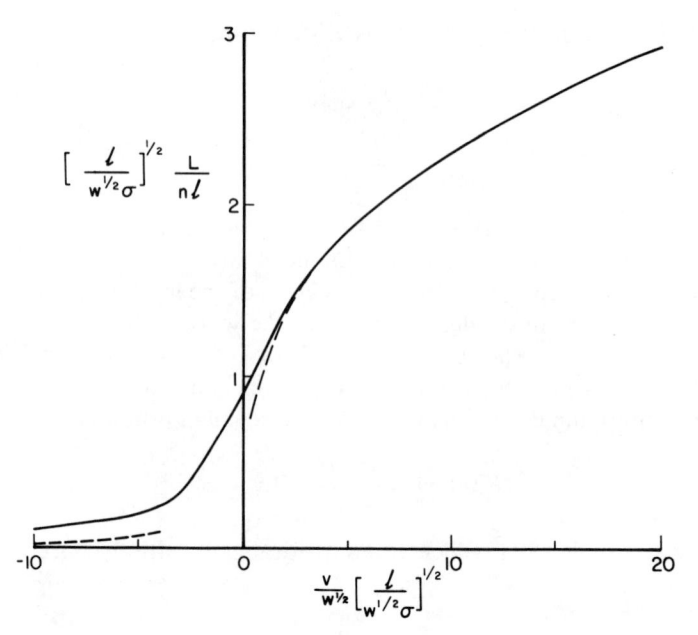

FIG. 27. Effect of solvent quality and graft density on layer thickness for terminally anchored chains from Eq. (112).

corresponding to the asymptotes in Fig. 27, provide additional insight. Note that $L \sim nl$ in all cases since the individual chains are highly stretched, but the layer thickness decreases with decreasing solvent quality or graft density. The sensitivity to the graft density, σ, increases as the solvent quality decreases. In the poor solvent case with $\bar{v} \ll 1$, the theory holds only for $-n^{1/2}w\sigma/vl \gg 1$.

The segment density profiles obtained from Eq. (110) have the limiting forms

$$\bar{v} \gg 1, \quad \frac{\rho l}{\sigma} \sim \left(\frac{9\pi}{32}\right)^{1/3} \left(\frac{vl^2}{w^{3/2}\sigma^2}\right)^{1/2} \left(1 - \frac{x^2}{L^2}\right)$$

$$\bar{v} \sim 0, \quad \frac{\rho l}{\sigma} \sim \left(\frac{2l}{w^{1/2}\sigma}\right)^{1/2} \left(1 - \frac{x^2}{L^2}\right)^{1/2} \quad (114)$$

$$\bar{v} \ll 1, \quad \frac{\rho l}{\sigma} \sim -2\pi \frac{vl}{w\sigma}$$

(Fig. 28). Clearly the profile becomes flatter with decreasing solvent quality. Note, however, that the self-consistent potential and the end-segment distributions still satisfy Eqs. (109) and (110).

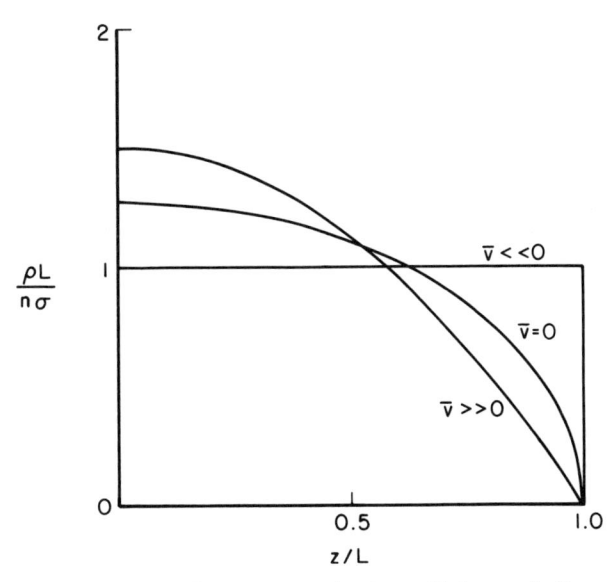

FIG. 28. Effect of solvent quality on segment density profile for terminally anchored chains from Eq. (114).

The predicted layer thicknesses resemble qualitatively those obtained for a step-function profile postulated by Alexander (1977) and de Gennes (1975). This mean-field approach approximates the free energy of the layer by

$$\frac{A}{kT} = \frac{3}{2}\left(\frac{L^2}{nl^2} + \frac{nl^2}{L^2} - 2\right) + \frac{nv\rho}{2} + \frac{nw\rho^2}{6}, \tag{115}$$

with constant segment density $\rho = n\sigma/L$. The bracketed term accounts for the elastic energy associated with compressing or extending the chains from the relaxed configuration, and the second and third terms represent the effects of two- and three-body segment–segment interactions. The equilibrium layer thickness then corresponds to the free energy for which the chemical potential of the solvent in the layer equals that in the bulk, or

$$\frac{\partial A}{\partial \rho} = -\frac{L^2}{n\sigma}\frac{\partial A}{\partial L} = 0.$$

Thus, the equilibrium layer thickness must satisfy

$$\left(\frac{L}{n^{1/2}l}\right)^3 - \left(1 + \frac{\phi_p^2}{9}\right)\frac{n^{1/2}l}{L} = \frac{n^{3/2}\sigma v}{6l}. \tag{116}$$

The dimensionless surface density, $\phi_p = n\sigma w^{1/2}/l$, is the volume fraction of segments that would result from collapsing the chains into a layer of thickness l. The second parameter, $n^{3/2}\sigma v/l$, is the ratio of the excluded volume per chain $(n^2 v)$ to the volume occupied by the chain at ideal conditions $(n^{1/2}l/\sigma)$.

The limiting forms for the layer thickness, e.g.,

$$L \sim \left(\frac{v\sigma}{6l}\right)^{1/3} nl, \quad \frac{n^{3/2}\sigma v}{l} \gg 1,$$

$$L \sim \left(\frac{w^{1/2}\sigma}{3l}\right)^{1/2} nl, \quad \frac{n^{3/2}\sigma v}{l} \sim 0 \quad \text{and} \quad \phi_p \gg 1, \quad (117)$$

$$L \sim -\frac{2}{3}\left(\frac{w\sigma}{lv}\right) nl, \quad \frac{n^{3/2}\sigma v}{l} \ll 1,$$

differ only in the numerical factors from Eq. (113). Thus, chains become highly stretched whenever either $n^{3/2}\sigma v/l$ or ϕ_p is large.

B. Interactions Between Layers

When two surfaces approach, the attached polymer layers first interact at separations h on the order of twice the layer thickness. The polymer–polymer and polymer–surface interactions alter the free energy by changing the volume available to an individual chain. For $v > 0$, interpenetration or compression of the two layers reduces the available volume and, hence, the number of configurations. The associated increase in free energy produces a repulsive force. However, with negative excluded volume, interpenetration increases the volume available per chain, thereby decreasing the free energy and causing an attraction. In both cases, reducing the separation to less than one layer thickness constrains the chains sufficiently to produce a strong repulsion.

The interaction potential equals the change in free energy per unit area, which is expressed according to Eq. (64) or alternatively as (Helfand, 1975a)

$$V = -2\sigma l^2 \ln W(h) - l^2 \int_0^h \left(\frac{1}{2}v\rho^2 + \frac{1}{3}w\rho^3\right) dz - \frac{l^2 A_\infty}{kT}, \quad (118)$$

with

$$W(h) = \int_0^h G(z, l/2, n)\, dz$$

and

$$A_\infty = \sigma kT \ln\left(2\pi \frac{n}{3}\right) + 2\sigma A_0\left(\frac{n^{3/2}\sigma v}{l}, \phi_p\right).$$

A_∞ is the free energy of the isolated layers with the first term comprising the ideal result and A_0 accounting for the effects of segment–segment interactions.

For ideal layers, solution of the SCF equation yields a small separation limit,

$$V(h) = \sigma l^2 \left[\frac{\pi^2 nl}{3h^2} + \ln\left(\frac{27h^2}{8\pi nl^2}\right) \right]. \tag{119}$$

The divergence of the potential as h^{-2} for $h \to 0$ arises from the compression of chains between the surfaces.

For dense layers in good solvents, the asymptotic theory of Milner et al. (1988) applies, but with $L = h/2$. Density profiles are truncated according to

$$\rho(x) = \frac{2n\sigma}{h} + \frac{\pi^2}{24v}\left[\left(\frac{h}{2nl}\right)^2 - 3\left(\frac{x}{nl}\right)^2\right], \tag{120}$$

indicating that chains in the interacting layers do not interpenetrate, but are simply compressed. The interaction potential then follows, albeit indirectly, from Eq. (118), the density profile, the form of the self-consistent potential, and the end-segment density as

$$V(h) = n\sigma l^2 \left(\frac{\pi^2}{12}\right)^{1/3} \left(\frac{v\sigma}{l}\right)^{2/3} \left[\frac{2L}{h} + \left(\frac{h}{2L}\right)^2 - \frac{1}{5}\left(\frac{h}{2L}\right)^5 - \frac{9}{5}\right]. \tag{121}$$

For slight compression, this simplifies to

$$V(h) \sim 3n\sigma l^2 \left(\frac{\pi^2}{12}\right)^{1/3} \left(\frac{v\sigma}{l}\right)^{2/3} \left(1 - \frac{h}{2L}\right)^3. \tag{122}$$

In the same limit, the mean-field approach based on a step function profile yields

$$V(h) \sim 3^{4/3} n\sigma l^2 \left(\frac{v\sigma}{2l}\right)^{2/3} \left(1 - \frac{h}{2L}\right)^2. \tag{123}$$

Thus, the approaches concur on the scaling of the potential, $n\sigma(v\sigma/l)^{2/3}$, and the separation, $nl(v\sigma/l)^{1/3}$, and predict a strong repulsion for $h < 2L$; they differ, however, in the form of the separation dependence. Analogous treatments for semidilute concentrations within the layers differ only modestly from the preceding results (Milner, 1989).

The asymptotic approach has not yet been extended to interactions between stretched layers in theta and poor solvents, which control the incipient flocculation of polymerically-stabilized colloidal dispersions. Application of the mean-field theory to poor solvents produces an attractive minimum only for $-n^{1/2}v/w^{1/2} > n\sigma w^{1/2}/l$, when layers begin to interpenetrate rather than simply compress as in good solvents (Russel *et al.*, 1989). This means denser layers provide greater stability, i.e., they require considerably worse than theta conditions for flocculation.

These potentials for interactions between flat plates provide the basis for addressing interactions between spheres of radius $a \gg L$ through the Derjaguin approximation

$$\frac{\Phi}{kT} = \frac{\pi a}{l^2} \int_{r-2a}^{\infty} V(h) \, dh.$$

For thicker layers, curvature effects become significant and must be treated explicitly (Witten and Pincus, 1986).

C. Experimental Results

Taunton *et al.* (1988) have measured forces between polystyrene layers in toluene, which are anchored to mica surfaces by small zwitterionic head

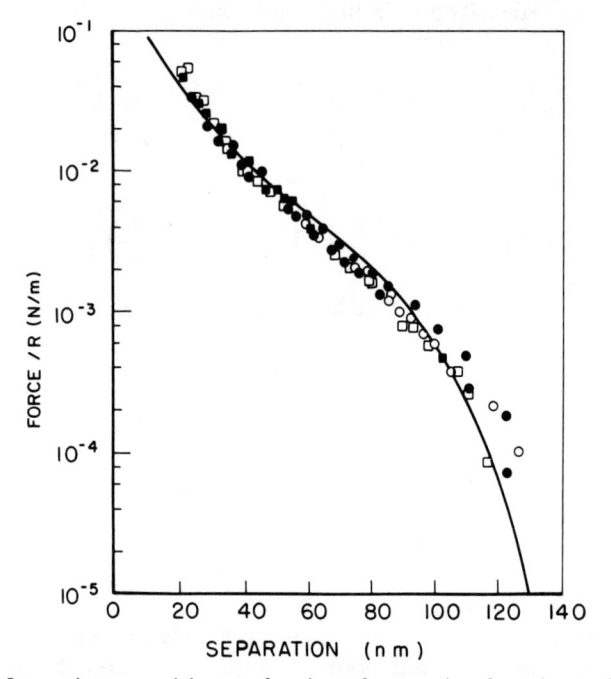

FIG. 29. Interaction potential as a function of separation for mica surfaces bearing terminally-anchored polystyrene chains with $M = 131$ kg/mol in toluene: Data points are taken from Taunton *et al.* (1988); the curve is from Eq. (121), with $n = 265$, $l = 1.46$ nm, $2L = 140$ nm, $nl^2\sigma = 22$, and $v/l^3 = 0.06$.

groups. That the polystyrene chain does not adsorb and the anchoring by the polar group is irreversible were demonstrated in independent measurements, so the situation approximates that assumed for the theory.

The interaction in toluene at room temperature, a good solvent, is strongly repulsive at separations less than twice the apparent layer thickness (Fig. 29). Polystyrene has statistical segments with a molecular weight of 0.494 kg/mol and a length of 1.46 nm, indicating $n = 265$, $nl = 398$ nm, and $\langle r^2 \rangle^{1/2} = 24$ nm. Thus, the apparent layer thickness of 60–70 nm detected in the experiment suggests highly stretched chains, presumably due to high graft densities. To test this quantitatively requires estimating σ and v/l^3 from the data. The solid curve in Fig. 29 results from choosing $L = 70$ nm and the prefactor in Eq. (121) as 1.14×10^{-3} N/m. With the good solvent limit for L from Eq. (113) this leads to $v/l^3 = 0.06$ and $nl^2\sigma = 22$, which are reasonable values.

Other experiments with block copolymers (Hadziioannou et al., 1986; Tirrell et al., 1987; Patel et al., 1988; Ansarifar and Luckham, 1988) provide similar results in good solvents and strong, but shorter range repulsions at theta conditions. The behavior in poor solvents, i.e., the delineation between attractive and repulsive potentials, remains to be resolved.

V. Nonadsorbing Polymer

A. Depletion Layers

Dissolved, nonadsorbing polymer molecules must sacrifice configurational degrees of freedom to approach to surface closer than about one coil radius, an energetically unfavorable process in dilute solutions. Indeed, experiments confirm that a layer of thickness $L \sim \langle r^2 \rangle^{1/2}$ adjacent to any surface is depleted of segments (Aussere et al., 1986).

The consequences for suspended particles can be understood from either a mechanical or a thermodynamic standpoint. A particle immersed in a polymer solution experiences an osmotic pressure acting normal to its surface. For an isolated particle, the integral of the pressure over the entire surface nets zero force. But when the depletion layers of two particles overlap, polymer will be excluded from a portion of the gap (Fig. 30). Consequently, the pressure due to the polymer solution becomes unbalanced, resulting in an attraction. The same conclusion follows from consideration of the Helmholtz free-energy. Overlap of the depletion layers reduces the total volume depleted of polymer, thereby diluting the bulk solution and decreasing the free energy.

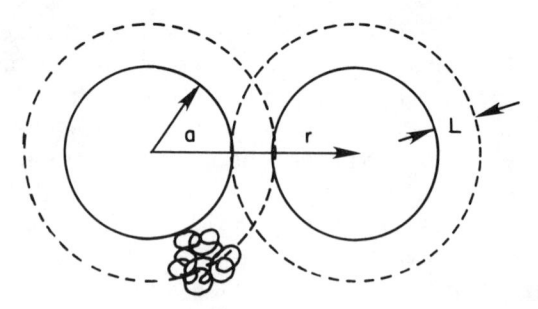

FIG. 30. Interaction between spheres of radius a at separation r showing overlap of depletion layers of thickness L.

The original geometrical analysis of Asakura and Oosawa (1954, 1958), generalized by Vrij (1976) and others, neglects the internal degrees of freedom of the polymer molecules to obtain simple, useful expressions for the interaction potential. The SCF theory reviewed here (Joanny et al., 1979) demonstrates the validity of the simpler approaches.

The configuration density function G, governed by Eq. (31) with $G = 0$ at $z = 0, h$, is now related to the segment density through Eq. (92) with σ replaced by ρ, the average density of segments within the gap, such that

$$\int_0^h \rho(z)\,dz = h\rho. \tag{124}$$

Equilibrium with a bulk solution at segment density ρ_b requires the chemical potential for chains in the gap to equal that for the bulk, or

$$\rho = \rho_b W \exp\left(nv\rho_b + \frac{\partial \ln W}{\partial \ln \rho} + \frac{\partial}{\partial \rho}\int_v \tfrac{1}{2} v\rho^2 \frac{dV}{V}\right). \tag{125}$$

The solution for G, obtained by assuming a separable solution of the form in Eq. (82) (Gerber and Moore, 1977) to the nonlinear Eq. (31) and requiring $g = 0$ at $z = 0, h$ and

$$\int_0^h g^2(z)\,dz = 1, \tag{126}$$

can be expressed as a Jacobian elliptic function and then used to evaluate W and $\rho(z)$. For sufficiently small separations, $\rho(z) \sim 2\rho \sin^2(\pi z/h)$ and

$$W \sim \exp\left(-\frac{\pi^2 n l^2}{6h^2}\right), \tag{127}$$

so that application of the equilibrium condition in Eq. (125) determines

$$\rho \sim \rho_b \exp\left(nv\rho_b - \frac{\pi^2 n l^2}{6h^2}\right). \tag{128}$$

Thus, the segment density within the gap effectively vanishes when $h < \pi n^{1/2} l/[6(1 + nv\rho_b)]^{1/2}$. For theta conditions, this corresponds to $h < \pi \langle r^2 \rangle^{1/2}/6$ as suggested above. For good solvents with $nv\rho_b > 1$, the interactions in the bulk favor configurational changes near the interface, so chains remain in the gap until $h < \pi l (6v\rho_b)^{1/2} = \xi$. These predictions pertain to the dilute and concentrated regimes in Fig. 3; the concentration dependence of the depletion layer thickness in the semidilute regime differs in the dependence of ξ on ρ_b (Joanny et al., 1979).

In both theta and good solvents, the exclusion of chains from the gap when $h < 2L$, with L as the depletion layer thickness, allows the interaction potential to be written as

$$V(h) = \frac{l^2}{kT}[A(h) - A(2L)]$$

$$= -(2L - h)\frac{\Pi l^2}{kT}, \tag{129}$$

with Π as the osmotic pressure of the bulk solution so that the force

$$F = -\frac{kT}{l^2}\frac{dV}{dh}$$

$$= -\Pi,$$

conforms to the original results of Asakura and Oosawa (1954, 1958). Combining the mean-field expression for Π with Eq. (129), evaluated at $h = 0$, yields the minimum in the potential energy at contact as

$$V_{\min} = -\frac{\rho_b}{n}\left(1 + \frac{v\rho_b}{2}\right)\frac{\pi l^3}{[6(1/n + v\rho_b)]^{1/2}}. \tag{130}$$

Comparison of these potentials with those for the terminally anchored chains shows the interaction to be relatively weak. For example, experiments with polystyrene in cyclohexane, which does not adsorb on mica, yielded no detectable forces between mica surfaces because of the polymer (Luckham and Klein, 1985). Indeed, estimates of the potential from Eq. (130) at the experimental conditions fall several orders of magnitude below the detection limit for the instrument.

The simple form of the interaction potential,

$$V = -\frac{\Pi V_{\text{excl}}}{kT}, \tag{131}$$

with V_{excl} as the volume of overlap between the two depletion layers, permits straightforward generalization to other geometries (Vrij, 1976). For equal spheres of radius a at center-to-center separation r,

$$\Phi(r) = -\frac{4\pi}{3}(a+L)^3\left(1 - \frac{3r}{4(a+L)} + \frac{r^3}{16(a+L)^3}\right)\Pi \tag{132}$$

decreases monotonically from zero at $r = 2(a + L)$ to a minimum (Fig. 31),

$$\frac{\Phi_{\min}}{kT} = -\frac{4\pi L^3}{3}\left(1 + \frac{3a}{2L}\right)\frac{\Pi}{kT}, \tag{133}$$

at $r = 2a$. In ideal solutions with $a/L \gg 1$

$$\frac{\Phi_{\min}}{kt} \sim \frac{2\pi}{3} a l^2 \rho_b$$

becomes $0(1)$ for $\rho_b l^3 \sim l/a \ll 1$. Hence, while not measurable between macroscopic surfaces, the interaction is sufficiently strong to affect the phase behavior of colloidal dispersions.

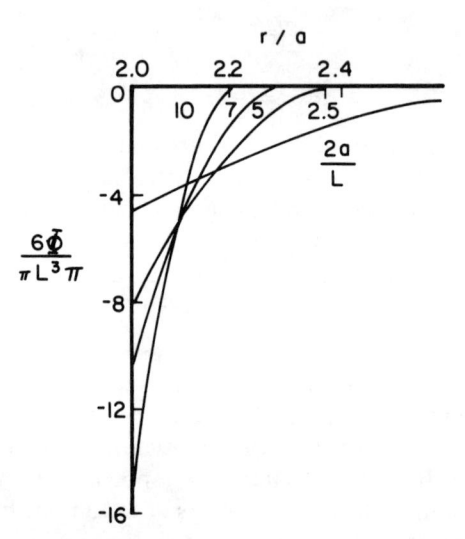

FIG. 31. Interaction potential between spheres of radius a with depletion layers of thickness L from Eq. (132) (Gast et al., 1983b).

B. Effect of Terminally Anchored Chains

The presence of grafted polymer chains with molecular weight n_* alters both the depletion of the free polymer and the thickness of the grafted layer as described by de Gennes (1980) and Gast and Leibler (1986). Within the mean-field approach, the penetration of solution polymer into the grafted layer and the effect on the layer thickness are described by expressing the free-energy densities of the layer and the bulk solution as

$$\frac{A_*}{kT} = \frac{3\sigma}{2L}\frac{L^2}{n_*l^2} + \frac{1}{2}v(\rho_* + \rho)^2 + \frac{\rho}{n}\ln(l^3\rho)$$

and (134)

$$\frac{A}{kT} = \frac{1}{2}v\rho_b^2 + \frac{\rho_b}{n}\ln(l^3\rho_b),$$

with ρ_* and ρ as the segment densities for attached and free chains within the layer. Recall that the segment density and layer thickness for the grafted chains are related by $\rho_* L = n_* \sigma$. Equilibrium requires equal chemical potentials for the solvent and free polymer in the two regions or, equivalently, equal exchange potentials for the free polymer, i.e., $\partial A_*/\partial \rho = \partial A/\partial \rho_b$, and equal osmotic pressures, i.e.,

$$\rho_* \partial A_*/\partial \rho_* + \rho \partial A_*/\partial \rho - A_* = \rho_b \partial A/\partial \rho_b - A.$$

The form of the resulting equations identifies the three distinct regimes depicted in Fig. 32:

$$\rho_b < \left(\frac{6\sigma^2}{l^2 v}\right)^{1/3}, \quad \rho \sim \rho_b \exp\left[-\left(\frac{6v^2\sigma^2}{l^2}\right)^{1/3} n\right]$$

$$L \sim \left(\frac{v\sigma}{6l}\right)^{1/3} n_* l \left[1 - l^3 \rho_b \left(\frac{v}{6l^3}\right)^{1/3} (l^2\sigma)^{2/3}\right]$$

$$\left(\frac{6\sigma^2}{l^2 v}\right)^{1/3} < \rho_b < (3n)^{1/2}\frac{\sigma}{l}, \quad \rho \sim \rho_b \exp\left[-3n\left(\frac{\sigma}{l\rho_b}\right)^2\right]$$

$$L \sim \frac{n_*\sigma}{\rho_b}\left(1 + \frac{3n^2}{vl^2\rho_b^3}\right)^{-1} \quad (135)$$

$$(3n)^{1/2}\frac{\sigma}{l} < \rho_b, \quad \rho \sim \rho_b\left[1 - (3n)^{1/2}\frac{\sigma}{l\rho_b}\right]$$

$$L \sim \frac{n_* l}{(3n)^{1/2}}.$$

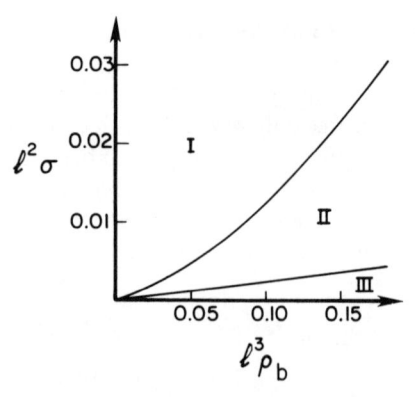

FIG. 32. Delineation of regimes for interaction between free polymer at bulk concentration $l^3\rho_b$ and terminally anchored chains of graft density $l^2\sigma$ for $n = 5000$, $p = 500$, and $v/l^3 = 1$ (Gast and Leibler, 1986): I, Negligible interpenetration and layer thickness unaffected by free polymer; II, slight interpenetration, but layer significantly compressed by free polymer; III, complete interpenetration and relaxed layer.

Thus, at low free-polymer concentrations, the penetration into the layer is exponentially small and the resulting contraction is negligible. However, when the bulk concentration exceeds the initial segment density within the layer, the bulk osmotic pressure causes significant contraction of the layer, though penetration remains small. Ultimately, at high concentrations, the layer relaxes to a thickness that is independent of both ρ_b and σ. The first two regimes are asserted to be the most important except, perhaps, for very low molecular-weight polymer in solution.

Detailed analysis, via the SCF theory, of the free polymer between two flat surfaces bearing grafted layers with constant segment density yielded profiles for ρ varying smoothly from ρ_b in the bulk to the asymptotic values cited above within the grafted layer (Gast and Leibler, 1986). The transition occurs within a region of dimension $l[1/6v\rho_*)^{1/2} + (6v\rho_b)^{1/2}]$ at the interface between the grafted layer and the solution. Thus, the depletion layer penetrates somewhat into the grafted layer, permitting the segment density at the interface to depart from zero. For example, in region II (Fig. 32).

$$n \sim \frac{1}{2}\rho_b \left(\frac{\rho_b}{\rho_*}\right)^{1/2}$$

at large surface-to-surface separations; the factor $(\rho_b/\rho_*)^{1/2}$ represents the ratio of the correlation length within the layer to that in the solution.

This diminution of the depletion layer reduces the attractive potential;

thus,

$$V_{min} = -\frac{\pi}{2}l^3 \rho_b \left(\frac{v\rho_b}{6}\right)^{1/2} \left\{1 - \left[\frac{\rho_b}{(6\sigma^2/vl^2)^{1/3}}\right]^{1/2}\right\}, \qquad (136)$$

in accordance with the earlier, qualitative arguments of de Gennes (1980). The prefactor in Eq. (136) corresponds to the attractive minimum in the absence of the grafted layer, but now $V_{min} \to 0$ when $\rho_b \to (6\sigma^2/vl^2)^{1/3}$. One interesting feature of this result is that increasing the graft density increases the magnitude of the attraction, contrary to intuition. At separations less than $2L$, the interaction of the grafted layers produces the same strong repulsion described in the previous section.

VI. Macroscopic Consequences

A. Phase Separations

For colloidal systems, the homogeneous dispersed state becomes unstable when the attractive minimum in the interparticle potential exceeds a few kT. For $-5 \leq \Phi_{min}/kT \leq -2$, dispersions of submicron particles separate on a reasonable time scale, i.e., a few days, into two coexisting equilibrium phases. With stronger attractions, equilibration becomes quite slow, requiring up to six months or a year. The nature and compositions of the phases depend on the range and magnitude of the attractive potential.

Phase separations induced by dissolved nonadsorbing polymer were recognized and exploited in the 1930s to concentrate, or cream, rubber latices (Napper, 1983, Sect. 15.2) and reappeared in the 1970s in the formulation of coatings (e.g., Sperry et al., 1981). Experiments with a variety of dispersions—polymer latices, silica spheres, microemulsions, rodlike DNA—in aqueous solutions of nonionic polymers or polyelectrolytes and both ideal and nonideal solutions in organic solvents have clearly established the coexistence of equilibrium phases, one dense and one dilute in particles, above a critical polymer concentration that correlates with $-\Phi_{min}/kT = 2 - 3$. Consequently, the critical polymer concentration decreases with increasing molecular weight or particle size in accordance with Eq. (132) (e.g., Cowell et al., 1978; Vincent et al., 1980; Sperry et al., 1981; de Hek and Vrij, 1981).

Subsequently, more detailed studies have revealed two modes of phase behavior depending on the ratio of the sizes of the particle and the polymer (Gast et al., 1983a,b, 1986; Sperry, 1984; Vincent, 1987). For example, electrostatically stabilized polystyrene latices in aqueous solutions of dextran exhibit a simple fluid–solid transition for $a/R_g = 6.9$, but they exhibit a more

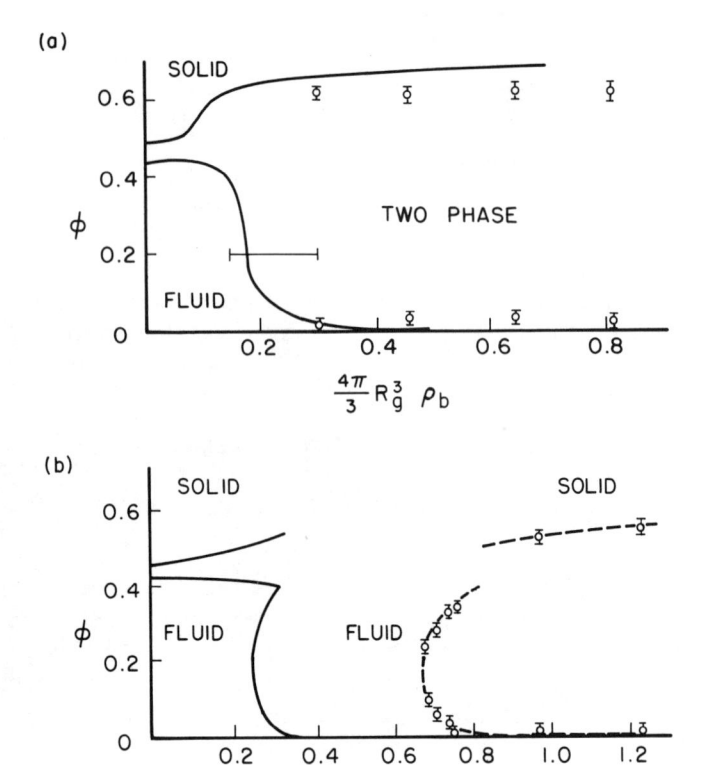

FIG. 33. Equilibrium phase behavior of polystyrene latices of radius a at volume fraction ϕ containing soluble dextran of radius of gyration R_g at fluid concentration ρ_b (Patel and Russel, 1989a). (a) $a/R_g = 6.9$: ⌀, measured compositions; ⊢⊣, bounds on transition; —, predicted phase boundaries. (b) $a/R_g = 1.9$: ⌀, measured compositions; ---, experimental phase boundary; —, predicted phase boundary.

complex phase diagram with a fluid–fluid envelope, critical and triple points, and a fluid–solid region for $a/R_g = 1.9$ (Fig. 33) (Patel and Russel, 1989a). The solid phases display iridescence, characteristic of crystalline order, and the expected mechanical responses, i.e., yield stresses and finite low-frequency elasticity; the fluid phases are opaque and flow as Newtonian liquids in the low shear limit (Patel and Russel, 1989b).

At modest concentrations of free polymer, all dispersions behave generally as described above, independent of the mechanism stabilizing the particles in the absence of polymer, e.g., grafted polymer chains or electrostatic repulsion. However, with increasing polymer concentration, the correlation length and, hence, the depletion layer thickness decrease, so the range and nature of the repulsion ultimately become important.

For particles stabilized with grafted polymer, generalization of the interaction potential in Eq. (136) to spheres via the Derjaguin approximation provides (Russel et al., 1989)

$$\frac{\Phi_{min}}{kT} = -\frac{\pi^3}{24} a l^2 \rho_b \left[1 - 0.74 \left(\frac{lv^{1/2}\rho_b^{3/2}}{\sigma} \right)^{1/3} \right]^2. \tag{137}$$

The second term in brackets reflects the penetration of free polymer into the grafted layer with increasing concentration and limits the maximum attraction to

$$\frac{\Phi_{min}}{kT} = -\frac{\pi^3}{108} \frac{a}{l} \left(\frac{l^7 \sigma^2}{v} \right)^{1/3} \tag{138}$$

at $\rho_b = \frac{9}{4}(\sigma/lv^{1/2})^{2/3}$. Any smaller value can be achieved with either of two concentrations, e.g., for $a/l \gg 1$

$$l^3 \rho_b = -\frac{24}{\pi^3} \frac{l}{a} \frac{\Phi_{min}}{kT}$$

or (139)

$$l^3 \rho_b = \frac{16}{9} \left(\frac{l^7 \sigma^2}{v} \right)^{1/3} \left[1 - \left(\frac{6l}{\pi^3 a} \right)^{1/2} \left(-\frac{\Phi_{min}}{kT} \right)^{1/2} \left(\frac{l^7 \sigma^2}{v} \right)^{1/6} \right].$$

Edwards et al. (1984) performed experiments with silica spheres bearing a layer of grafted polystyrene chains in a solution of somewhat higher molecular-weight polystyrene in toluene. The stability boundaries depicted in Fig. 34 indicate aggregation, detected optically, over a finite range of polymer concentrations at volume fractions of particles ϕ exceeding a minimum value. Since the equilibrium ratio of doublets to singlets in a dilute dispersion should vary as

$$\phi \exp\left(-\frac{\Phi_{min}}{kT} \right),$$

the attraction required to produce significant aggregation must increase with decreasing volume fraction. Thus, the potential described above leads to a stability boundary of exactly the form observed with the minimum volume fraction proportional to

$$\exp\left[-\frac{\pi^3}{108} \frac{a}{l} \left(\frac{l^7 \sigma^2}{v} \right)^{1/3} \right]$$

and the two branches of the curve related to the concentrations in Eq. (139).

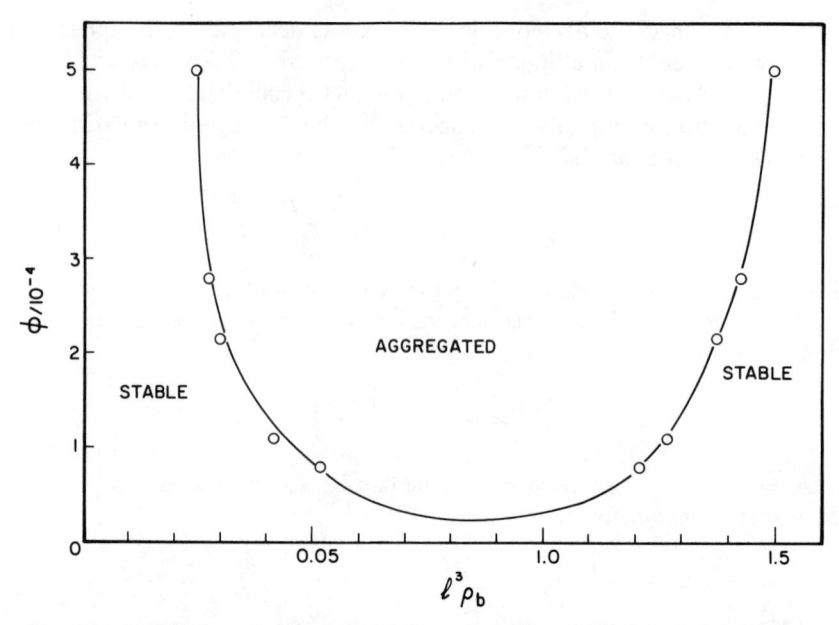

FIG. 34. Stability boundary for silica spheres ($a = 115$ nm) bearing grafted polystyrene chains ($M = 7.5$ kg/mol) in solution of polystyrene ($M = 31$ kg/mol) in toluene (Edwards et al., 1984).

These results demonstrate that nonadsorbing polymer can induce phase separations in colloidal systems with the nature of the phases depending primarily on the ratio of the particle and polymer sizes. Since the strength of the attraction is not necessarily a monotonic function of the polymer concentration, e.g., because of penetration of the free polymer into a grafted layer, both destabilization and restabilization are possible.

B. Polymeric Stabilization

The interaction potentials described in previous sections for adsorbing homopolymer and terminally anchored layers in good solvents clearly indicate the ability of polymers to stabilize colloidal dispersions against flocculation due to van der Waals dispersion forces. Indeed, the practice preceeded the analyses by centuries in some cases and decades in others, since the use of adsorbing polymers dates to ancient times, and block copolymer stabilizers emerged from industrial laboratories in the 1960s (Napper, 1983).

Polymeric stabilization offers several advantages over the electrostatic primary alternative:

(1) The absence of electroviscous effects,
(2) Stability in nonaqueous solvents with low dielectric constants and low surface charge densities,
(3) Robust stability over long periods of time and at high volume fractions, and
(4) Reversible flocculation if the solvent quality degrades.

Stability clearly requires a good solvent for the stabilizer and a repulsion of sufficiently long range to mask the dispersion forces, i.e., a minimum layer thickness. In addition, for adsorbing homopolymers, the interaction potentials are entirely repulsive only for strong adsorption at full coverage, Even then, though, the repulsion appears only when the long time-scale for desorption constrains polymer within the gap during interactions between particles. Otherwise, the greater free energy of the macromolecules squeezed between the surfaces drives a slow desorption, thereby reducing the repulsion. Indeed a slow flocculation, or aging, is possible in dispersions sufficiently concentrated to maintain the layers in contact. Hence, homopolymers are not the optimum stabilizers.

With chains anchored to the surface, either by a chemical grafting or an insoluble block, good solvent conditions always produce a repulsion. Consequently, copolymers, e.g., diblock, comb, or graft, tend to comprise the most effective stabilizers. Direct grafting to the particle is feasible but requires chemistry specific to the particle (e.g., Green et al., 1987). Advances in synthetic polymer chemistry continue to increase the types of polymers available for this application (e.g., Reiss et al., 1987).

The two primary features of the phenomena are the layer thickness necessary to provide stability and the conditions at which the dispersions flocculate. The first can be quantified by generalizing the potential for terminally anchored chains to interactions between spheres via the Derjaguin approximation, adding the attractive dispersion potential, and then assessing the layer thickness necessary to maintain $-\Phi_{min}/kT < 1 - 2$. To illustrate this, consider the small overlap limit of Eq. (122), which transforms into

$$\frac{\Phi}{kT} = \frac{\pi a}{l^2} \int_{r-2a}^{2L} V(h)\, dh$$
$$= \left(\frac{9\pi^5}{32}\right)^{1/3} n\sigma a L \left(\frac{v\sigma}{l}\right)^{2/3} \left(1 - \frac{r-2a}{2L}\right)^4 \qquad (140)$$

for equal spheres of radius a. A suitable approximation for the retarded dispersion potential between spheres with relatively thick polymer layers in nonaqueous liquids is (Russel, et al., 1989, sect. 9.2)

$$\frac{\Phi}{kT} = -\frac{2^{1/2}A_{\text{eff}}}{3\pi kT} \frac{\lambda}{n_0^2(n_0^{-2} + n_0^2)^{1/2}} \frac{a}{(r-2a)^2},\qquad(141)$$

with A_{eff} as the nonretarded limit of the effective Hamaker constant, λ as the wavelength characteristic of the ultraviolet relaxation of the liquid, and n_0 and $\bar{n}_0$ as the low frequency limits of the refractive indices at visible wavelengths for the liquid and the particles, respectively.

Modest calculations demonstrate that the minimum occurs at

$$\frac{r-2a}{2L} \sim 1$$

if $\lambda/[L^3 n(v^2\sigma^5/l^2)^{1/3}] \ll 1$, which is invariably true. This leads to a minimum layer thickness of

$$\frac{L_{\min}}{a} \sim \left[\frac{\lambda A_{\text{eff}}}{6\pi 2^{1/2} akT n_0^2(n_0^{-2}+n_0^2)^{1/2}}\right]^{1/2}\qquad(142)$$

and, along with Eq. (112), determines the requisite molecular weight. Since $A_{\text{eff}}/kT \sim 1-10$, n_0 and $\tilde{n}_0 \sim 1.2$, and $\lambda \sim 0.1$ µm, the minimum layer thickness falls in the range $(0.05-0.20)a$.

The condition for incipient flocculation, i.e., the delineation between stable and unstable dispersions, generally corresponds to the theta condition for the stabilizer in solution. A large body of experimental data gathered by Napper and others (e.g., Napper, 1983) first established this correlation, and the theoretical predictions for the interaction potentials agree provided $nl^2\sigma \leq 1$.

At higher surface coverages, the mean-field treatment noted in Section IV indicates that worse than theta conditions are required before an attractive minimum appears in the potential. This phenomena has been observed and termed "enhanced steric stabilization" by Napper (1983). Recent experiments by Edwards et al. (1984) clearly show that grafting PDMS chains onto silica particles at $nl^2\sigma \sim 12-18$ extends the stable range substantially beyond theta conditions.

C. Polymeric Flocculation

Water soluble polymers serve widely as flocculants, particularly in the water treatment, paper making, and minerals industries. The objective generally is to destabilize dispersions and cause large, strong, and compact flocs to form quickly. The flocs are often removed subsequently by sedimentation or flotation. Polymers perform efficiently without introducing the salt needed to suppress the electrostatic forces largely responsible for the stability of the dispersions (Rose and St. John, 1985). Polyelectrolytes, bearing multiple charges along their backbones or on side groups, provide enhanced solubility and promote adsorption onto particles of opposite charge. High molecular weights, $10-10^4$ kg/mol, tend to be preferred.

A combination of electrostatic and kinetic effects complicate the process, with no single mechanism appearing to control all systems of interest. Even interpretations based on the equilibrium interparticle potential present different possibilities (e.g., Vincent, 1974; Rose and St. John, 1985):

(1) *Charge neutralization.* Polyelectrolytes of opposite charge adsorb electrostatically, thereby neutralizing the surface charge and permitting dispersion forces to drive flocculation. Then the conditions for optimum flocculation should correlate with the point of zero charge for the particles.

(2) *Bridging flocculation.* Partial coverage of adsorbing nonionic polymer generally results in a strong attractive minimum in the pair potential, since individual macromolecules span the gap and attach to both particles. This implies optimum conditions for flocculation at surface coverages of about one half monolayer per surface.

Some suggestions concerning the conditions that specific mechanisms control emerge from the analyses of Section III. For example, lower molecular weight polymers adsorb with relatively flat conformations, making bridging across the electrical double layers unlikely; thus, charge neutralization might dominate in that limit. For higher molecular weights at moderate concentrations, the depth and position of the attractive minimum depend on the surface coverage (Fig. 17); hence, significant bridging should occur if the range exceeds twice the double layer thickness before the depth of the minimum becomes negligible, i.e., restabilization occurs. Conversion of the minimum, V_{min}, between flat surfaces to equal spheres via the Derjaguin approximation produces an attraction of roughly $aR_g V_{min}/l^2$. At the conditions of Fig. 25, $-aR_g V_{min}/l^2 \sim 1.5 - 14$ (with $a/l \sim 10^2$) for polymer concentrations of $l^3 \rho_b < 10^{-3}$.

Experiments by Pelssers (1988) probe these issues for PEO adsorbed onto polystyrene at different surface coverages. At low molecular weights, the adsorption equilibrated under stable conditions before the ionic strength was raised to $10^{-2} M$ to initiate flocculation. The growth of small flocs was then detected with a novel scattering instrument. A distinct onset of flocculation appeared at a critical polymer dose (Fig. 35), presumably corresponding to the surface coverage at which the range of the attraction due to bridging exceeded twice the double layer thickness. Restabilization was not observed because of the limited surface coverage achievable with such low molecular weights.

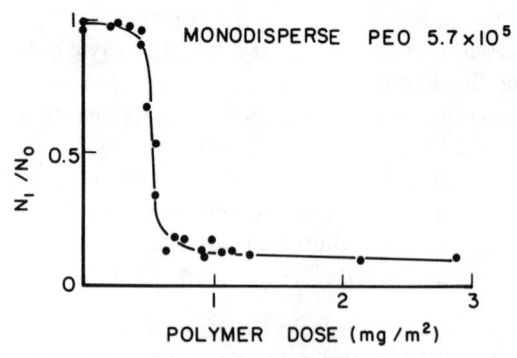

FIG. 35. Effect of the amount of PEO added on the fraction of individual polystyrene latex particles remaining (N_1/N_0) after 1.5 hours (Pelssers, 1988): $M = 57$ kg/mol, $2a = 0.7$ μm, $\phi = 4 \times 10^{-4}$, $[KNO_3] = 10^{-2}$ M.

Kinetic measurements established binary collisions between unstable particles and flocs as the controlling process.

For higher molecular weights, however, the phenomena differs qualitatively in the dependence on time, initial number density of particles, and polymer dosage. Then the rate of polymer reconformation during the adsorption relative to the rate of particle collisions becomes a key consideration (Gregory, 1987). The data in Fig. 36 (Pelssers, 1988) reveal a transition from encounters of particles bearing equilibrium-adsorbed layers, at low particle number densities and long time-scales for collisions, to bridging by adsorbed polymer with dimensions comparable to the coil in solution at higher number densities

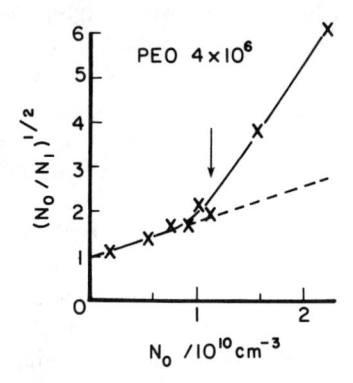

FIG. 36. Effect of the initial number density of polystyrene latices ($2a = 0.7$ μm) on the fraction of singlets remaining 1.5 hours after the addition of PEO at 0.34 mg/m². The dashed line characterizes the doublet formation with relaxed chains and b indicates the transition to flocculation before reconformation (Pelssers, 1988).

and, hence, shorter collision times. Since the reconformation continues during the flocculation process, the rate decreases monotonically with time, in some cases arresting the process with flocs of finite size. Clearly the efficacy of higher molecular-weight polymers must arise at least in part from such kinetic effects.

For polyelectrolytes, analyses of equilibrium adsorption are only beginning to emerge. Muthukumar's (1987) treatment of a single polyelectrolyte chain interacting with a charged surface predicts adsorption with a mean square end-to-end distance of $O(\kappa^{-1})$ when

$$-\frac{lqq_s}{\varepsilon kT} > (l\kappa)^{11/5} n^{1/5}$$

with q, the charge per segment on the polymer; q_s the surface charge density; $\kappa^{-1} = (\varepsilon kT/e^2 I)^{1/2}$ (where I is ionic strength and e is electronic charge) the Debye length; and ε, the dielectric constant for the electrolyte solution. Hence, adsorption occurs only below a critical ionic strength. The consequences with respect to the mechanisms just cited and the interaction potentials between two surfaces remain to be explored.

Thus, the situation is currently clouded by an incomplete understanding of both the kinetic processes and the equilibrium behavior of polyelectrolytes. For specific systems, consistent correlations have been constructed, but a general understanding of the functioning of polymeric flocculants requires more extensive fundamental experiments and development of the equilibrium theory for polyelectrolytes.

D. Rheology

Several recent studies demonstrate convincingly the possibilities for adjusting the rheology of colloidal dispersions through the incorporation of polymer. Here we briefly review the effects of grafted polymer, adsorbing homopolymer, and nonadsorbing polymer. The literature abounds with other and more complicated phenomena.

A grafted layer of polymer of thickness L increases the effective size of a colloidal particle. In general, dispersions of these particles in good solvents behave as non-Newtonian fluids with low and high shear limiting relative viscosities (η_0 and η_∞), and a dimensionless critical stress ($a^3\sigma_c/kT$) that depends on the effective volume fraction $\phi_{\text{eff}} = (1 + L/a)^3 \phi$. The viscosities diverge at volume fractions ϕ_{m0} and $\phi_{m\infty}$, respectively, with $\phi_{m0} < \phi_{m\infty}$; for $\phi_{m0} < \phi_{\text{eff}} < \phi_{m\infty}$, the dispersions yield and flow as pseudoplastic solids.

The data of Mewis *et al.* (1989) for poly(methyl methacrylate) spheres stabilized by poly(12-hydroxy stearic acid) and dispersed in decalin conform

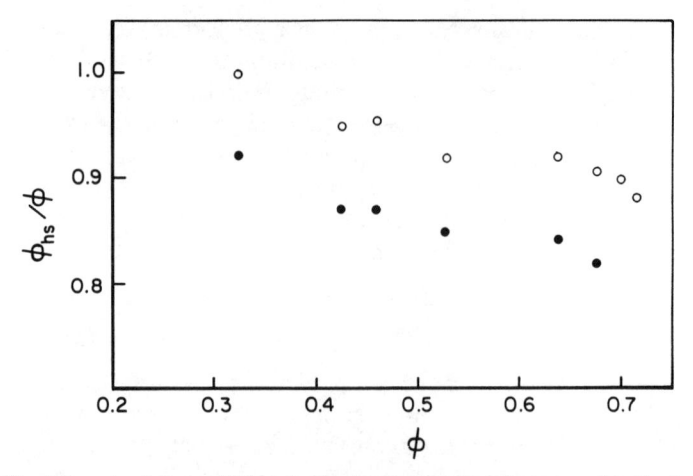

FIG. 37. The ratio of the equivalent hard sphere volume fraction φ_{hs} to the effective volume fraction ϕ based on the measured intrinsic viscosity as a function of ϕ for poly(methyl methacrylate) spheres with grafted poly(12-hydroxy stearic acid) layers such that $a/L = 4.7$ (Mewis et al., 1989). Open and closed circles correspond to the low and high shear limits of suspension viscosity.

to results for hard spheres for $\phi_{eff} \leq 0.4$, although the critical stress is somewhat smaller. For $\phi_{eff} \geq 0.4$, however, the softness of the repulsion becomes evident. At rest, packing constraints cause some interpenetration of the layers and at high shear rates viscous forces drive the particles even closer together. Consequently, the ratio of the equivalent hard-sphere volume fraction, obtained by comparing the measured viscosities with those for hard spheres, to ϕ_{eff} decreases with increasing ϕ_{eff} and Pe (Fig. 37).

The grafted layer also affects two other features of the rheology. First, thicker polymer layers enhance the elasticity due to the longer range of the repulsion relative to the hard core size. Thus, samples formulated at $\phi_{eff} \geq \phi_{m0}$ possess easily measurable static elastic moduli. Second, the softer repulsion apparently suppresses the shear thickening observed at high volume fractions for the harder particles, in accordance with earlier measurements by Willey and Macosko (1978).

The situation with adsorbing homopolymer differs significantly because of ability of the polymer to either stabilize or bridge colloidal particles, depending on the characteristics of the polymer and particles in the quiescent state. In addition, shear can either induce bridging or breakup flocs formed by bridging. Experiments by Otsubo and co-workers (Otsubo and Umeya, 1983, 1984; Otsubo, 1986; Otsubo and Watanabe, 1987, 1988) with silica spheres in solutions of poly(acrylamide) in glycerin–water mixtures illustrate the effects of particle size and volume fraction, polymer molecular weight and concen-

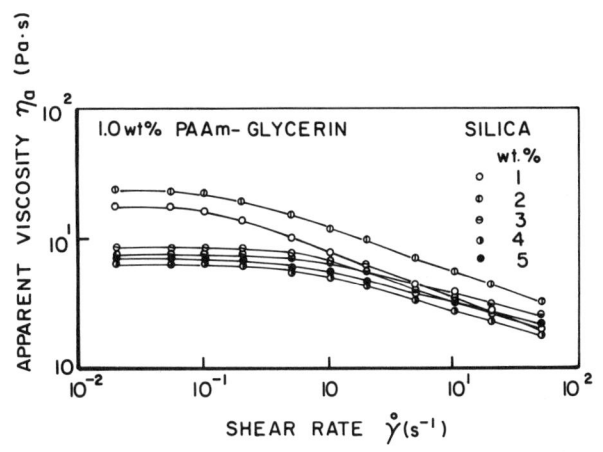

FIG. 38. Apparent viscosity of various dispersions of silica spheres ($2a = 20$ nm) in solutions of 1 wt% polyacrylamide ($M_w = 2 \times 10^3$ kg/mol) in glycerin (Otsubo and Umeya, 1984).

tration, solvent quality, and shear rate. Figs. 38 and 39 reflect the most striking features of the behavior.

In glycerin, the dispersion appears to be stable and behaves as a shear thinning fluid. However, the viscosity does not vary monotonically with silica concentration, instead decreasing initially before increasing at higher concentrations (Fig. 38). The phenomena reflect offsetting effects of adsorption, the reduction in the solution viscosity due to depletion of polymer

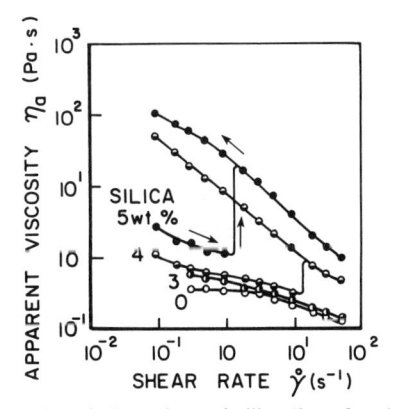

FIG. 39. Apparent viscosity of dispersions of silica ($2a = 2$ nm) in 0.5 wt% solutions of polyacrylamide ($M_w = 5.5 \times 10^3$ kg/mol) in 50/50 glycerin–water mixtures (Otsubo and Watanabe, 1987). Different symbols correspond to different weight percentages of silica.

and the increased hydrodynamic volume of the particles. The particles adsorb ~ 1.5 mg/m^2 of polymer on a surface area of 0.13 m^2/mg, so ~ 5 wt% silica substantially depletes the solution. At higher silica concentrations, interactions among the coated particles increase the viscosity, suggesting a layer thickness of ~ 10 nm, and lead to a dilatancy at higher shear rates. Thus, the coated particles behave as stable spheres at all shear rates.

Better solvents, e.g., glycerin–water mixtures, enhance the possibility of bridging, perhaps by reducing the surface coverage. The rheological behavior then becomes quite sensitive to those parameters controlling the surface coverage: The molecular weight and concentration of the polymer, the size and concentration of the particles, and the solvent quality (Fig. 39). At high surface coverages, i.e., large particles, low silica concentrations, or high polymer molecular weights, the systems flow as relatively low viscosity, stable dispersions. But at rather low coverages, arising for smaller particles, higher silica concentrations, or lower molecular weights, bridging produces very viscous, pseudoplastic materials. At intermediate coverages, an irreversible shear-induced transition occurs, transforming a low viscosity dispersion into a paste at a critical stress. While these interpretations are speculative, the dramatic effects warrant further study, perhaps with better characterized polymer and particles and more extensive measurements of the adsorption and the adsorbed layer thickness.

With nonadsorbing polymer, rheological effects of similar magnitude accompany the phase transitions described earlier (Patel and Russel, 1989a,b). Since macroscopic phase separation takes weeks or months, rheological measurements performed within a few days on samples formulated within the two-phase region, with $-\Phi_{\min}/kT \sim 2 - 20$, detect a metastable structure that changes little over time. The systems respond as flocculated dispersions, but the microstructure recovers relatively quickly to a reproducible rest state after shear. Hence these weakly flocculated dispersions are quite tractable materials.

The steady shear viscosities in Fig. 40 demonstrate the primary features of the phenomena:

(1) A transition from fluid to solid in the mechanical response accompanying the frustrated fluid–solid phase transition,

(2) A yield stress that increases smoothly but dramatically with increasing polymer concentration, and

(3) A pronounced shear thinning of the fluid phase near or within the fluid–fluid region.

Linear viscoelastic measurements detect a low frequency plateau in the shear modulus whenever the yield stress appears. The magnitudes of both correlate

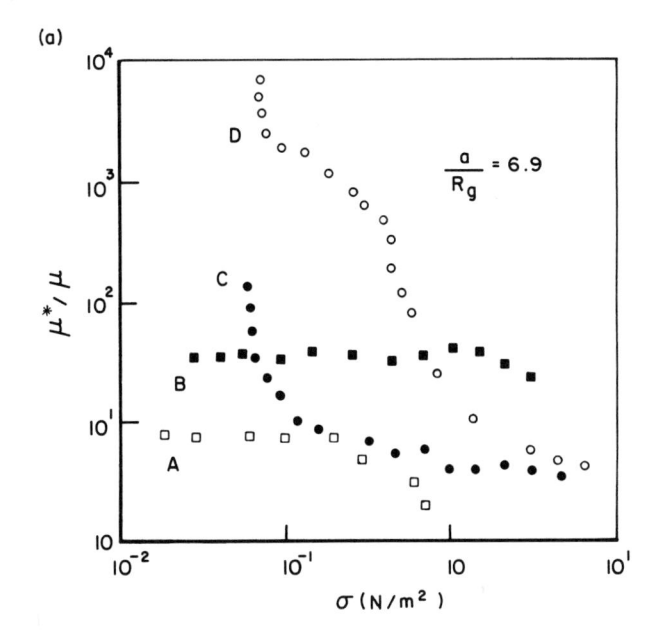

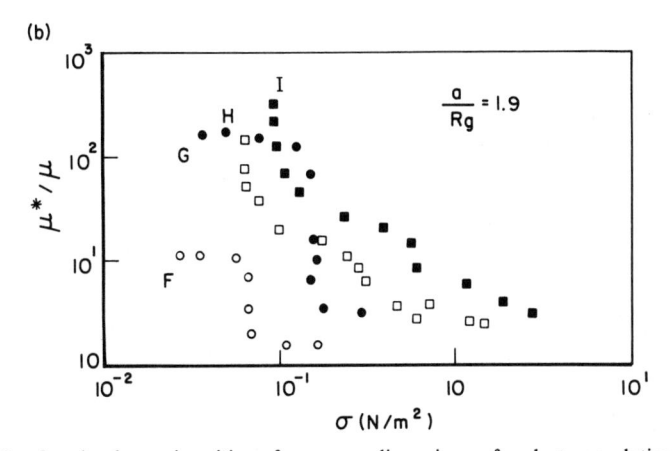

FIG. 40. Steady shear viscosities of aqueous dispersions of polystyrene latices in nonadsorbing dextran solutions (Patel and Russel, 1989b): (a) $a/r_g = 6.9$, $\phi = 0.20$. A, single phase, $4\pi R_g^3/3\rho_b = 0.15$; B, two-phase, $4\pi R_g^3/3\rho_b = 0.30$; C, two-phase, $4\pi R_g^3/3\rho_b = 0.45$; D, two-phase, $4\pi R_g^3/3\rho_b = 0.65$. (b) $a/R_g = 1.9$, $\phi = 0.10$. F, single phase, $4\pi R_g^3/\rho_b = 0.65$; G, fluid–fluid, $4\pi R_g^3/3\rho_b = 0.75$; H, fluid–solid, $4\pi R_g^3/3\rho_b = 0.95$; I, fluid–solid, $4\pi R_g^3/3\rho_b = 1.25$.

with characteristics of the interaction potential discussed in Section V, providing a convenient way of adjusting the rheology.

These examples serve to illustrate the ability of soluble polymer, interacting in a controlled fashion with colloidal particles, to transform both the equilibrium state and the mechanical properties of dispersions. All states are possible, from low viscosity fluids to pseudoplastic pastes with high yield stresses.

References

Alexander, S., *J. Phys. (Orsay, Fr.)* (1977).
Almog, Y., and Klein, J., *J. Colloid Interface Sci.* **106**, 33 (1985).
Anderson, J., and Kim, J. O., *J. Chem. Phys.* **86**, 5163 (1987).
Ansarifar, M. A., and Luckham, P. F., *Polymer* **29**, 329 (1988).
Asakura, S., and Oosawa, F., *J. Chem. Phys.* **22**, 1255 (1954).
Asakura, S., and Oosawa, F., *J. Polym. Sci.* **33**, 183 (1958).
Aussere, D., Hervet, H., and Rondalez, F., *Macromolecules* **19**, 85 (1986).
Auvray, L., and Cotton, J. P., *Macromolecules* **20**, 202 (1987).
Auvray, L., and de Gennes, P. G., *Europhys. Lett.* **2**, 647 (1986).
Baram, A., and Luban, M., *J. Phys. C* **12**, L659 (1979).
Bawendi, M. G., Freed, K. F., and Mohanty, U., *J. Chem. Phys.* **84**, 7036 (1986).
Baxter, R. J., *J. Chem. Phys.* **49**, 2770 (1969).
Billmeyer, F. W., "Textbook of Polymer Science." Wiley, New York, 1982.
Bovey, F. A., "Chain Structure and Conformation of Macromolecules." Academic Press. New York, 1982.
Brandrup, J., and Immergut, E. H., eds. *in* "Polymer Handbook," 2nd Ed. Wiley, New York, 1975.
Brebner, K. I., Chahal, R. S., and St-Pierre, L. E., *Polymer* **21**, 533 (1980a).
Brebner, K. I., Brown, G. R., Chahal, R. S., and St-Pierre, L. E., *Polymer* **22**, 56 (1980b).
Cahn, J., *J. Chem. Phys.* **66**, 3667 (1977).
Cahn, J., and Hilliard, J. E., *J. Chem. Phys.* **28**, 258 (1958).
Carnahan, N. F., and Starling K. E., *J, Chem. Phys.* **51**, 625 (1969).
Cebula, D., Thomas, R. K., Harris, N. M., Tabony, J., and White, J. W., *Faraday Discuss Chem. Soc.* **65**, 76 (1978).
Chandrasekhar, S., *Rev. Mod. Phys.* **15**, 1 (1943).
Charmet, J. C., and de Gennes, P. G., *J. Opt. Soc. Am.* **73**, 1773 (1983).
Cohen Stuart, M. A., Ph. D. Thesis, Agric. Univ., Wageningen, Netherlands, 1980.
Cohen Stuart, M. A., Fleer, G. J., and Scheutjens, J. M. H. M., *J. Colloid Interface Sci.* **97**, 515 (1984a).
Cohen Stuart, M. A., Fleer, G. J., and Scheutjens, J. M. H. M., *J. Colloid Interface Sci.* **97**, 526 (1984b).
Cohen Stuart, M. A., Waajen, F. H. F. H., Cosgrove, T., Vincent, B., and Crowley, T. L., *Macromolecules* **17**, 1825 (1984c).
Cohen Stuart, M. A., Cosgrove, T., and Vincent, B., *Adv. Colloid Interface Sci.* **24**, 143 (1986).
Cosgrove, T., Vincent, B., Crowley, T. L., and Cohen Stuart, M. A., *ACS Symp. Ser.* No. 240, 147 (1984).
Cosgrove, T., Heath, T. G., Ryan, K., and Crowley, T. L., *Macromolecules* **20**, 2879 (1987a).
Cosgrove, T., Heath, T. G., Ryan, K., and van Lent, B., *Polym. Commun.* **28**, 64 (1987b).

Cotton, J. P., Nierlich, M., Boue, F., Daoud, M., Farnoux, B., Jannink, G., Duplessix, G., and Picot, C., *J. Chem. Phys.* **65,** 1101 (1976).
Cowell, C., Li-In-On, R., and Vincent, B., *J. C. S. Faraday I* **74,** 337 (1978).
Croxton, C. A., *J. Phys. A* **12,** 2497 (1978).
Daoud, M., and Jannink, G., *J. Physc. (Orsay, Fr.)* **37,** 973 (1976).
Daoud, M., Cotton, J. P., Farnoux, B., Jannink, G., Sarma, G., Benoit. H., Duplessix, G., Picot, C., and de Gennes, P. G., *Macromolecules* **8,** 804 (1975).
de Gennes, P. G., *Rep. Prog. Phys.* **32,** 187 (1969).
de Gennes, P. G., *Phys. Lett. A* **38,** 339 (1972).
de Gennes, P. G., *J. Phys. (Orsay, Fr.)* **36,** L55 (1975).
de Gennes, P. G., "Scaling Concepts in Polymer Physics." Cornell Univ. Press, Ithaca, New York, 1979.
de Gennes, P. G., *Macromolecules* **13,** 1069 (1980).
de Gennes, P. G., *Macromolecules* **14,** 1637 (1981).
de Gennes, P. G., *Macromolecules* **15,** 492 (1982).
de Gennes, P. G., *Adv. Colloid Interface Sci.* **27,** 189 (1987).
de Gennes, P. G., and Pincus, P., *J. Phys., Lett. (Orsay, Fr.)* **44,** L241 (1983).
de Hek, H., and Vrij. A., *J. Colloid Interface Sci.* **84,** 409 (1981).
Des Cloizeaux, J., *J. Phys. (Orsay, Fr.)* **36,** 281 (1975).
Des Cloizeaux, J., *J. Phys., Lett. (Orsay, Fr.)* **41,** L151 (1980a).
Des Cloizeaux, J., *J. Phys. (Orsay, Fr.)* **41,** 749 (1980b).
Des Cloizeaux, J., *J. Phys. (Orsay, Fr.)* 761 (1980c).
Des Cloizeaux, J., *J. Phys. (Orsay, Fr.)* **42,** 635 (1981).
Dickman, R., and Hall, C. K., *J. Chem. Phys.* **85,** 4108 (1986).
DiMarzio, E. A., *J. Chem. Phys.* **42,** 2101 (1965).
DiMarzio, E. A., and McCrackin, F. L., *J. Chem. Phys.* **43,** 539 (1965).
DiMarzio, E. A., and Rubin, R. J., *J. Chem. Phys.* **55,** 4318 (1971).
Dolan, A. K., and Edwards, S. F., *Proc. R. Soc. London, Ser.* **A337,** 509 (1974).
Dolan, A. K., and Edwards, S. F., *Proc. R. Soc. London, Ser.* **A343,** 427 (1975).
Domb, C., *J. Chem. Phys.* **38,** 2957 (1963).
Domb, C., *Adv. Chem. Phys.* **15,** 229 (1969).
Douglas, J. F., Wang, S. Q., and Freed, K. F., *Macromolecules* **19,** 2207 (1986).
Douglas, J. F., Wang, S. Q., and Freed, K. F., *Macromolecules* **20,** 543 (1987).
Drude, P., *Ann. Phys.* **272,** 532 (1889a).
Drude, P., *Ann. Phys.* **272,** 865 (1889b).
Drude, P., *Ann. Phys.* **275,** 481 (1890).
Edwards, J., Lenon, S., Toussaint, A. F., and Vincent, B., *ACS Symp. Ser.* No. 240, 281 (1984).
Edwards, S. F., *Proc. Phys. Soc.* **85,** 613 (1965).
Edwards, S. F., *Proc. Phys. Soc.* **88,** 265 (1966).
Edwards, S. F., *J. Phys. A* **8,** 1670 (1975).
Edwards, S. F., and Jeffers, E. J., *J. C. S. Faraday II* **75,** 1020 (1979).
Edwards, S. F., and Singh, P., *J. C. S. Faraday II* **75,** 1001 (1979).
Eirich, F. R., *J. Colloid Interface Sci.* **58,** 423 (1977).
Eisenriegler, E., *J. Chem. Phys.* **79,** 1052 (1983).
Eisenriegler, E., *J. Chem. Phys.* **81,** 4666 (1984).
Eisenriegler, E., *J. Chem. Phys.* **82,** 1032 (1985).
Eisenriegler, E., Kremer, K., and Binder, K., *J. Chem. Phys.* **77,** 6296 (1982).
Evans, E., *Macromolecules,* **22,** 2277 (1989).
Fleer, G. J., *in* "Reagents in Mineral Technology" (P. Somasundaran and B. M. Moudgil, eds.). Dekker, New York, 1988.

Fleer, G. J., and Lyklema, J., *in* "Adsorption from Solution at the Solid/Liquid Interface" (G. D. Parfitt and C. H. Rochester, eds.). Academic Press, New York, 1983.
Fleer, G. J., and Scheutjens, J. M. H. M., *J. Colloid Interface Sci.* **16,** 341 (1982a).
Fleer, G. J., and Scheutjens, J. M. H. M., *J. Colloid Interface Sci.* **16,** 361 (1982b).
Fleer, G. J., and Scheutjens, J. M. H. M., *J. Colloid Interface Sci.* **111,** 504 (1986).
Fleer, G. J., Cohen Stuart, M. A., and Scheutjens, J. M. H. M., *Colloids Surf.* **31,** 1 (1988).
Flory, P. J., *J. Chem. Phys.* **10,** 51 (1942).
Flory, P. J., "Principles of Polymer Chemistry." Cornell Univ. Press, Ithaca, New York, 1953.
Flory, P. J., "Statistical Mechanics of Chain Molecules." Wiley (Interscience), New York, 1969.
Flory, P. J., *Discuss Faraday Soc.* **49,** 7 (1970).
Forsman, W. C., ed., "Polymers in Solution." Plenum, New York, 1986.
Freed, K. F., *Adv. Chem. Phys.* **22,** 1 (1972).
Freed, K. F., *J. Phys. A* **18,** 871 (1985).
Freed, K. F., "Renormalization Group Theory of Macromolecules." Wiley (Interscience), New York, 1987.
Frisch, H. L., *J. Phys. Chem.* **59,** 633 (1955).
Frisch, H. L., and Simha, R., *J. Phys. Chem.* **58,** 507 (1954).
Frisch, H. L., and Simha, R., *J. Chem. Phys.* **24,** 652 (1956).
Frisch, H. L., and Simha, R., *J. Chem. Phys.* **27,** 702 (1957).
Frisch, H. L., Simha, R., and Eirich, F. R., *J. Chem. Phys.* **21,** 365 (1953).
Gast, A. P., *Proc. NATO ASI Ser.*, Ser. E, Strasbourg, France, 1990.
Gast. A. P., and Leibler, L., *Macromolecules* **19,** 686 (1986).
Gast, A. P., Hall, C. K., and Russel, W. B., *J. Colloid Interface Sci.* **96,** 251 (1983a).
Gast, A. P., Hall, C. K., and Russel, W. B., *Faraday Discuss* **76,** 189 (1983b).
Gast, A. P., Russel, W. B., and Hall, C. K., *J. Colloid Interface Sci.* **109,** 161 (1986).
Gerber, P. R., and Moore, M. A., *Macromolecules* **10,** 476 (1977).
Green, M., Kramer, T., Parish, M., Fox, J., Lalanandham, R., Rhine, W., Barclay, S., Calvert, P., and Bowen, H. K., *Adv. Ceram.* **21,** 449 (1987).
Gregory, J., *Colloids Surf.* **31,** 231 (1987).
Hadziioannou, G., Patel, S., Granick, S., and Tirrell, M., *J. Am. Chem. Soc.* **108,** 2869 (1986).
Halperin, A., *J. Phys. (Orsay, Fr.)* **49,** 547 (1988).
Helfand, E., *J. Chem. Phys.* **63,** 2192 (1975a).
Helfand, E., *Macromolecules* **8,** 307 (1975b).
Helfand, E., *J. Chem. Phys.* **62,** 999 (1975c).
Helfand, E., *Macromolecules* **9,** 307 (1976).
Helfand, E., and Sapse, A. M., *J. Chem. Phys.* **62,** 1327 (1975).
Helfand, E., and Sapse, A. M., *J. Polym. Sci., Polym. Symp.* **54,** 289 (1976).
Helfand, E., and Wasserman, Z., *Macromolecules* **9,** 879 (1976).
Hesselink, F. T., *J. Phys. Chem.* **73,** 3488 (1969).
Hesselink, F. T., *J. Phys. Chem.* **75,** 65 (1971).
Hoeve, C. A. J., *J. Chem. Phys.* **43,** 3007 (1965).
Hoeve, C. A. J., *J. Chem. Phys.* **44,** 1505 (1966).
Hoeve, C. A. J., DiMarzio, E. A., and Peyser, P., *J. Chem. Phys.* **42,** 2558 (1965).
Hong, K. M., and Noolandi, J., *Macromolecules* **14,** 727 (1981).
Huggins, M. L., *Ann. N.Y. Acad. Sci.* **41,** 1 (1942).
Ishinabe, T., *J. Chem. Phys.* **76,** 5589 (1982).
Israelachvili, J. N., Tandon, R. K., and White, L. R., *J. Colloid Interface Sci.* **78,** 430 (1980).
Israelachvili, J. N., Tirrell, M., Klein, J., and Almog, Y., *Macromolecules* **17,** 204 (1984).
Jenckel, E., and Rumbach, B., *Z. Elektrochem.* **55,** 612 (1951).
Joanny, J. J., Leibler, L., and de Gennes, P. G., *J. Polym. Sci., Polym. Phys. Ed.* **17,** 1073 (1979).
Jones, I. S., and Richmond, P. J., *J. C. S. Faraday II* **73,** 1062 (1977).

Kato, T., Nakamura, K., Kawaguchi, M., and Takahashi, A., *Polym. J.* **13**, 1037 (1981).
Kawaguchi, M., Hayakawa, K., and Takahashi, A., *Polym. J.* **12**, 265 (1980).
Kawaguchi, M., Hayakawa, K., and Takahashi, A., *Macromolecules* **16**, 631 (1983).
Killmann, E., and Bergmann, M., *Colloid Polym. Sci.* **263**, 381 (1985).
Killmann, E., Eisenlauer, J., and Korn, M., *J. Polym. Sci., Polym. Symp.* **61**, 413 (1977).
Klein, J., *Adv. Colloid Interface Sci.* **16**, 101 (1982).
Klein, J., *J. C. S. Faraday I* **79**, 99 (1983).
Klein, J., and Luckham, P. F., *Nature (London)* **308**, 836 (1984a).
Klein, J., and Luckham, P. F., *Macromolecules* **17**, 1041 (1984b).
Kosmas, M., and Freed, K. F., *J. Chem. Phys.* **69**, 3647 (1978).
Kurata, M., "Thermodynamics of Polymer Solutions." Harwood, New York, 1982.
Levine, S., Thomlinson, M. M., and Robinson, R., *Faraday Discuss Chem. Soc.* **65**, 202 (1978).
Lifshitz, I. M., Grosberg, A. Y., and Khokhlov, A. R., *Rev. Mod. Phys.* **50**, 683 (1978).
Luckham, P. F., and Klein, J., *Macromolecules* **18**, 721 (1985).
Ma, S., "Modern Theory of Critical Phenomena." Benjamin, Reading, Massachusetts, 1976.
McCrackin, F. L., and Colson, J. P., *NBS Misc. Publ. (U.S.)* No. 256, 61 (1964).
McQuarrie, D. A., "Statistical Mechanics." Harper, New York, 1976.
Marques, C., Joanny, J. F., and Leibler, L., *Macromolecules* **21**, 1051 (1988).
Mewis, J., Frith, W., Strivens, T. A., and Russel, W. B., *AIChE J.* **35**, 415 (1989).
Milner, S. T., Witten, T. A., and Cates, M. E., *Europhys. Lett.* **5**, 413 (1988).
Milner, S. T., *Europhys. Lett.* **7**, 695 (1988).
Moore, M. A., *J. Phys. A* **10**, 305 (1977a).
Moore, M. A., *J. Phys. (Orsay, Fr.)* **38**, 265 (1977b).
Moore, M. A., and Al-Noaimi, G. F., *J. Phys. (Orsay, Fr.)* **39**, 1015 (1978).
Munch, M. R., and Gast, A. P., *Macromolecules* **21**, 1360, 1366 (1988).
Muthukumar, M., *J. Chem. Phys.* **86**, 7230 (1987).
Muthukumar, M., and Edwards, S. F., *J. Chem. Phys.* **76**, 2720 (1982).
Napper, D. H., "Polymeric Stabilization of Colloidal, Dispersions." Academic Press, New York, 1983.
Nemirovsky, A. M., and Freed, K. F., *J. Chem. Phys.* **83**, 4166 (1985).
Noda, I., Kato, N., Kitano, T., and Nagasawa, M., *Macromolecules* **14**, 668 (1981).
Ohta, T., and Oono, Y., *Phys. Lett. A* **89**, 460 (1982).
Okamoto, H., *J. Chem. Phys.* **64**, 2686 (1975).
Oono, Y., and Freed, K. F., *J. Chem. Phys.* **75**, 993 (1981).
Oono, Y., Ohta, T., and Freed, K. F., *J. Chem. Phys.* **74**, 6458 (1981).
Otsubo, Y., *J. Colloid Interface Sci.* **112**, 380 (1986).
Otsubo, Y., and Umeya, K., *J. Colloid Interface Sci.* **95**, 279 (1983).
Otsubo, Y., and Umeya, K., *J. Rheol.* **28**, 95 (1984).
Otsubo, Y., and Watanabe, K., *J. Non-Newtonian Fluid Mech.* **24**, 265 (1987).
Otsubo, Y., and Watanabe, K., *J. Colloid Interface Sci.* **122**, 346 (1988).
Patel, P. D., and Russel, W. B., *J. Colloid Interface Sci.* **131**, 192 (1989a).
Patel, P. D., and Russel, W. B., *J. Colloid Interface Sci.* **131**, 201 (1989b).
Patel, S., Tirrell, M., and Hadziioannou, G., *Colloids Surf.* **31**, 157 (1988).
Pelssers, E., Ph.D. Thesis, Agric. Univ., Wageningen, Netherlands, 1988.
Ploehn, H. J., Ph.D. Thesis, Princeton Univ., Princeton, New Jersey, 1988.
Ploehn, H. J., and Russel, W. B., *Macromolecules* **22**, 266 (1989).
Ploehn, H. J., Russel, W. B., and Hall, C. K., *Macromolecules* **21**, 1075 (1988).
Prigogine, I., "The Molecular Theory of Solutions." North-Holland Publ., Amsterdam, 1957.
Reiss, G., Bahadur, P., and Hurtvez, G., *Encycl. Polym. Sci. Eng.* **2**, 324 (1987).
Roe, R. J., *J. Chem. Phys.* **43**, 1591 (1965).
Roe, R. J., *J. Chem. Phys.* **44**, 4264 (1966).

Roe, R. J., *J. Chem. Phys.* **60,** 4192 (1974).
Rose, G. R., and St. John, M. R., *Encycl. Polym. Sci. Eng.* **7,** 211 (1985).
Rubin, R. J., *J. Chem. Phys.* **43,** 2392 (1965).
Russel, W. B., "The Dynamics of Colloidal, Systems." Univ. of Wisconsin Press, Madison, 1987.
Russel, W. B., Saville, D. A., and Schowalter, W. R., "Colloidal Dispersions." Cambridge Univ. Press, London, 1989.
Sanchez, I. C., and Lacombe, R. H., *Macromolecules* **11,** 1145 (1978).
Schaefer, D. W., *Polymer* **25,** 387 (1984).
Scheutjens, J. M. H. M., and Fleer, G. J., *J. Phys. Chem.* **83,** 1619 (1979).
Scheutjens, J. M. H. M., and Fleer, G. J., *J. Phys. Chem.* **84,** 178 (1980).
Scheutjens, J. M. H. M., and Fleer, G. J., *Macromolecules* **18,** 1882 (1985).
Scheutjens, J. M. H. M., Fleer, G. J., and Cohen Stuart, M. A., *Colloids Surf.* **21,** 285 (1986).
Silberberg, A., *J. Phys. Chem.* **66,** 1872 (1962a).
Silberberg, A., *J. Phys. Chem.* **66,** 1884 (1962b).
Silberberg, A., *J. Chem. Phys.* **46,** 1105 (1967).
Silberberg, A., *J. Chem. Phys.* **48,** 2835 (1968).
Silbergberg, A., *Pure Appl. Chem.* **26,** 583 (1971).
Simha, R., Frisch, H. L., and Eirich, F. R., *J. Phys. Chem.* **57,** 584 (1953).
Sperry, P. R., *J. Colloid Interface Sci.* **99,** 97 (1984).
Sperry, P. R., Hopfenberg, H. B., and Thomas, N. L., *J. Colloid Interface Sci.* **82,** 62 (1981).
Sperry, P. R., Thibeault, J. C., and Kostanek, E. C., *Adv. Coatings Sci. Technol.* **9,** 1 (1987).
Stanley, H. E., "Introduction to Phase Transitions and Critical Phenomena." Oxford Univ. Press (Clarendon), London and New York, 1971.
Stromberg, R. R., *In* "Treatise on Adhesion and Adhesives" (R. Patrick, ed.), Vol. 1, Chap. 3. Dekker, New York, 1967.
Stromberg, R. R., Passaglia, E., and Tutas, D. J., *J. Res. Natl. Bur. Stand., Sect. A* **67A,** 431 (1963).
Tabor, D., and Winterton, R. H. S., *Proc. R. Soc. London, Ser. A* **312,** 435 (1969).
Tadros, T. F., *In* "The Effect of Polymers on Dispersion Properties" (T. F. Tadros, ed.), pp. 1–38. Academic Press, London, 1982.
Tadros, T. F., *In* "Polymer Colloids" (R. Buscall, T. Corner, and J. F. Stageman, eds.), pp. 105–139. Elsevier, London, 1985.
Takahashi, A., and Kawaguchi, M., *Adv. Polym. Sci.* **46,** 1 (1982).
Takahashi, A., Kawaguchi, M., Hirota, H., and Kato, T., *Macromolecules* **13,** 884 (1980).
Taunton, H. J., Toprakcioglu, C., Fetters, L. J., and Klein, J., *Nature (London)* **332,** 712 (1988).
Tirrell, M., Patel, S., and Hadziioannou, G., *Proc. Natl. Acad. Sci. U.S.A.* **84,** 4725 (1987).
van der Beek, G. P., and Cohen Stuart, M. A., *J. Phys. (Orsay, Fr.)* **49,** 1449 (1988).
vander Linden, C., and van Leemput, R., *J. Interface Sci.* **67,** 48 (1978).
van Kampen, N. G., "Stochastic Processes in Physics and Chemistry." North-Holland Publ., Amsterdam, 1981.
Vincent, B., *Adv. Colloid Iinterface Sci.* **4,** 193 (1974).
Vincent, B., *Chem. Eng. Sci.* **42,** 779 (1987).
Vincent, B., Luckham, P. F., and Waite, F. A., *J. Colloid Interface Sci.* **73,** 508 (1980).
Vrij, A., *Pure Appl. Chem.* **48,** 471 (1976).
Weber, T. A., and Helfand, E., *Macromolecules* **9,** 311 (1976).
Weiss, G. H., and Rubin, R. J., *Adv. Chem. Phys.* **52,** 363 (1983).
Widow, B., *Physica A (Amsterdam)* **95A,** 1 (1979).
Willey, S. J., and Macosko, C. W., *J. Rheol.* **22,** 525 (1978).
Wiltzius, P., Haller, H. R., Cannell, D. S., and Schaefer, D. W., *Phys. Rev. Lett.* **51,** 1183 (1983).
Witten, T. A., and Pincus, P. A., *Macromolecules* **19,** 2509 (1986).
Yamakawa, H., "Modern Theory of Polymer Solutions." Harper, New York, 1971.
Ziman, J. M., "Models of Disorder." Cambridge Univ. Press, London, 1979.

INDEX

A

Accelerating rate calorimeter, reactive chemicals, 88–89
Adsorption isotherms
 Scheutjens-Fleer lattice model, 173–174
 self-consistent field theory, 188
Antisymmetric stress, 66
Autoignition temperatures, reactive chemicals, 90

B

Bridging flocculation, 217
Brinkman equation, 29–30
Brittleness, low temperature, 105
Bromotrifluoromethane, as extinguishing agent, 100
Brownian motion
 dilute suspensions, 26
 momentum tracer methods, 60, 62–63
 suspensions, 4
Bulk conductance, suspensions, 33
Buried tanks, chemical plant design, 81–81

C

Cell models, 5
 suspensions, 21–22
Centrosymmetric particles, suspensions, 45
Charge neutralization, 217
Chemical plant design, 73–74
 adequate redundancy of instrument and control systems, 110–114
 adequate space between process plants, tanks, and roads, 76–77
 avoiding catastrophic failure of engineering materials, 104–110

buffer zones, 77, 83
buried tanks, 81–81
clear responsibility for safety in design and operation, 75
continuous analog measurement, 111
critical instrument system, 113
diking for flammable liquids, 78–79
distance from residential areas, 83
dry quick-disconnect couplings, 124
emergency block valves, 118–120
enclosed flares, 103–104
expertise in, 74
fail-safe valves, 113–114
flexible or expansion joints, 127
gaskets, 121–123
glass and transparent devices, 126–127
hazard identification, 74–75
improving batch process productivity, 91–92
incinerating hazardous waste materials, 102–104
incorporation of emergency planning into, 76
in situ production and consumption of hazardous raw materials, 101–102
inventory reduction
 by changing mixing intensity, 92–93
 by changing process chemistry, 90–92
liquid storage, 78–79
liquified gas storage, 84–85
low inventory in distillation processes, 93–94
minimizing inventory in heat exchangers, 94–96
open structures, plants using flammable or combustible materials, 80–81
piping, 120–121
plastic and plastic lined pipe, 124–125
pressure relief systems, 114–118

229

Chemical plant design *(continued)*
  process and storage area construction, 83
  pumps suitable for hazardous service, 128–133
    deadhead pumps, 133
    metallurgy, 128
    seal-less pumps, 128–129
    types of seals, 129–132
  reactive chemicals testing, 87–90
  reactive hazard evaluations, 86
  redesigning obsolete plants, 84
  reduction of possibility of losses from dust explosions, 96–100
  review alternatives early in, 75–76
  safe and rapid isolation of piping systems or equipment, 118–120
  shutdown, 112–113
  single point signals, 111
  spring-loaded check valves, 124
  static mixer reactors, 93
  strong vessels to withstand maximum pressure of process upsets, 125–126
  substituting less hazardous materials in processes, transportation, and storage, 100–101
  total containment, 83–84
  triangular spacing between major process components, 77
  understanding reactive chemicals and systems involved, 86–90
  using minimum storage inventory of hazardous material, 78
  valves, 123–124
  worst case thinking, 86–87
Chemical potential, 176
Clusters, suspensions, 35
Coarse-graining, 148
Coefficient k_1, 23–24
Collision sphere, 40
Colloids, *see* Polymer–colloid–solvent mixtures
Combustible materials, open structures for plants, 80–81
Containment, dust explosions, 97
Continuity equation, averaged, suspensions, 7, 29
Continuum-mechanical principle of material-frame indifference, 67
Control systems, redundant, chemical plant design, 110–114
Copolymers, 138–139

Critical instrument system, 113
Cubic arrays, spatially periodic suspensions, 48–51

D

Deadheaded pumps, 133
Deflagration pressure containment, 125–126
Derjaguin approximation, 204, 213
Diblock copolymer, 138–139
DIERS, 114–115
Differential scanning calorimetry, reactive chemicals, 88
Differential thermal analysis, reactive chemicals, 88
Diisopropyl peroxydicarbonate, 102
Dilute suspensions, rheological models, 23–27
Distillation processes, low inventory, 93–94
Double-seal pumps, 129, 131
Drude equations, 189–190
Dry quick-disconnect couplings, 124
Dust explosions
  combustion process, 96
  containment, 97
  inerting, 98–99
  reactive chemicals, 90
  reducing possibility of losses, 96–100
  suppression systems, 99–100
  venting, 97–98

E

Emergency block valves, 118–120
Emergency planning, incorporation into chemical plant design, 76
Empirical models, suspensions, 19–21
Enclosed flares, 103–104
Engineering materials
  avoiding catastrophic failure, 104–110
  metals, 105–108
  nonmetals, 108–109
  plastic materials, 109–110
  Resista-Clad, 107–108
Environmental Protection Agency, buried tank requirements, 82
Evolution equation, 54–55
Expansion joints, 127
Explosion doors, 98
Explosion suppression systems, 99–100

Extinguishing agent, used in dust explosions, 100

F

Fail-safe valves, 113–114
Falling-ball suspension viscometry, 63–64
Fiberglass reinforced plastics, 110
Fickian diffusion tensors, 26–27
Fire-safe ball valve, 123
Flammable limits, reactive chemicals, 90
Flammable liquids, diking for, 78–79
Flammable materials, open structures for plants, 80–81
Flares, 103–104
Flash point, reactive chemicals, 90
Flixborough disaster, 78, 84
Flory–Huggins equation-of-state, 180–181
Flory–Huggins free energy of mixing, 152
Flory–Huggins mean-field theories, 152–153, 164, 191–193
Force–torque vector, 54
Fractal suspensions, 64–66
Free energy, randomly adsorbing homopolymers, 158

G

Gaskets, chemical plant design, 121–123
Glass devices, 126–127
Glass-lined steel, 108–109
Grand resistance matrix, 45, 51–52

H

Halon 1301, as extinguishing agent, 100
Hazardous materials
  in situ production and consumption, 101–102
  substitution in processes, transportation, and storage, 100–101
  using minimum storage inventory, 78
Hazardous waste materials, incineration, 102–104
Hazards, identification, 74–75
Heat exchangers, 94–96
Helmholtz free-energy, 155–156, 175–176
  self-consistent field theory, 191
Higee distillation process, 94
High pressure systems, 125

Homopolymer, *see also* Scheutjens–Fleer lattice model; Self-consistent field theory
  conformation in solution, 137–138
  randomly adsorbing, 157–197
    adsorbed amount and free energy, 165–167
    adsorbing boundary condition, 159
    adsorption energies and critical energy, 160
    apparent surface volume fraction, 165
    apparent viscosity, 220–222
    architecture, 157
    comparison with scattering measurements, 167–169
    configurations, 157–158, 159
    equilibrium and excluded volume, 161
    Flory–Huggins theory, 164
    free energy, 158, 163
    fundamental scaling prediction, 165
    individual ideal chains, 158–161
    interaction potentials, 166
    lattice models and matrix methods, 161–164
    matrix of transition probabilities, 162
    random walks, 159
    scaling and renormalization group theories, 164–169
    self-similarity, 164–165
    structure factors, 167–168
    surface tension, 165
    tail length, 160
Hydrodynamic force, suspensions, 8
Hydrodynamic thickness, 174–175, 191–192
Hydrogen embrittlement, metals, 106

I

Incineration, hazardous waste materials, 102–104
Inerting, dust explosions, 98–99
In situ manufacturing, 101–102
Institute for Emergency Relief Systems, 114–115
Instrument systems, redundant, chemical plant design, 110–114
Interaction potential
  nonadsorbing polymers, 207–208
  polymer–colloid–solvent mixtures, 215
  randomly adsorbing, polymers, 166
  Scheutjens–Fleer lattice model, 176–178

Interaction potential *(continued)*
  terminally anchored polymers, 202, 204
  versus gap half-width, 193–196
Internal spin field, 66
Inventory reduction
  by changing mixing intensity, 92–93
  by changing process chemistry, 90–92
  distillation processes, 93–94
  heat exchangers, 94–96
  storage, hazardous materials, 78

L

Lattice models, 161–164
Leaks, buried tanks, detection, 82
Leibniz packing, 64–65
Liouville equation, 15, 25
Liquid storage, design, 78–79
Liquified gases, storage, 84–85
Local gradient operator, 59
Low Reynolds number hydrodynamics, many-body problem, 10–13

M

Macroscopic stress, suspensions, 17
Macroscopic velocity gradient, 16
Magnetic fluids, vortex viscosity, 66–67
Many-body problem
  low Reynolds number hydrodynamics, 10–13
  statistical formulation, 13–15
Matrix method, 161–164
Mechanical energy dissipation rate, dilute suspensions, 23
Menger sponge, modified, 65–66
Metallurgy, pumps, 128
Metals, 105–108
  hydrogen embrittlement, 106
  low temperature brittleness, 105
  nitrate stress corrosion, 107
  stress-crack corrosion, 105–106
Method of induced forces, 12
Microrheology, suspensions, 20
Midland, MI explosion, 80–81
Momentum tracer methods, 57–64
  Brownian motion, 60, 62–63
  falling-ball suspension viscometry, 63–64
  local gradient operator, 59
  molecular diffusion, 60

no-slip boundary condition, 62
rigid-particle case, 62
second-order total moment, 60–61
steady-state unit cell equations, 61
Stokes equations, 59–60
suspension-scale kinematic viscosity, 58–69
tensor fields, 61
transport equation, 61
Momentum transport, suspensions, 34

N

Neuron-scattering experiments, homopolymer, randomly adsorbing, 167–169
Nitrate stress corrosion, metals, 107
Nonmetals, 108–109

O

Osmotic pressure
  scaling theory, 146–148
  self-consistent field theory, 192

P

Particle stress, suspensions, 8
Partitioned matrix relation, 9
Percolation models, 34–36
Percolation theory, 32–34
Phase function, suspensions, 28–29
Phase separations, polymer–colloid–solvent mixtures, 211–214
Phosgene, 100–101
Piping, chemical plant design, 120–121
Plastic lined pipe, 124–125
Plastic materials, 109–110
Plastic pipe, 125
Polyelectrolytes, 217
Polymer–colloid–solvent mixtures
  bridging flocculation, 217
  charge neutralization, 217
  copolymers, 138–139
  Derjaguin approximation, 213
  grafted layer, 219–220
  homopolymers, *see* Homopolymers
  initial number density of polystyrene lattices, 218
  interaction potential, 215

maximum attraction limit, 213
minimum layer thickness, 216
nonadsorbing polymer, *see* Polymers, nonadsorbing
PEO effects, 217–218
phase separations, 211–214
polymeric flocculation, 216–219
polymeric stabilization, 214–216
polymer properties, 140
polystyrene latices, 211–212
ratio of equivalent hard sphere volume fraction to effective volume fraction, 220
retarded dispersion potential between spheres, 215–216
rheology, 219–224
stability boundary, 213–214
terminally anchored, *see* Polymers, terminally anchored
thermodynamics, 140–156
  dilute limit, 142
  excluded volume parameter, 141
  Flory mean-field theories, 152–153
  higher-order interactions, 142
  polymer volume fraction, 143
  random-walk model, 141
  renormalization-group theories, 148–151
  scaling theory, 144–148
  self-consistent, mean-field theories, 153–156
  semidilute regime, 143–144
  temperature-concentration diagram, 143–144
universal models, 140–141
Polymeric flocculation, polymer–colloid–solvent mixtures, 216–219
Polymeric stabilization, polymer–colloid–solvent mixtures, 214–216
Polymers
  nonadsorbing, 205–211
    configuration density function, 205
    delineation of regimes, 209–210
    depletion layers, 205–208, 210–211
    interaction between spheres, 205–206
    interaction potential, 207–208
    minimum in potential energy, 207, 211
    overlap of depletion layers, 205–206
    rheological effects, 222
    terminally anchored chains, 209–211
  properties in solution, 142

terminally anchored, 197–205
  boundary condition, 198
  conformation, 197–198
  density profiles, 203
  Derjaguin approximation, 204
  end segment probability, 199
  experimental results, 204–205
  free energy, 201
  interaction potential, 202, 204
  interactions between layers, 202–204
  isolated layers, 197–202
  segment density profiles, 200–201
  solvent quality effect on thickness, 199–200
Poly(methyl methacrylate) spheres, 219–220
Polystyrene latices
  equilibrium phase behavior, 211–212
  initial number density, 218
  polymer–colloid–solvent mixtures, 211–212
  steady shear viscosities, 222–223
Pressure-relief systems
  ARC, 116
  chemical plant design, 114–118
  isolation of safety valves, 118
  two-phase flow, 115, 118–119
  VSP, 116

Q

Quarter-turn valves, 123–124

R

Random copolymer, 138–139
Random-walk model, 141
Reactive chemicals
  accelerating rate calorimeter, 88–89
  autoignition temperatures, 90
  differential scanning calorimetry, 88
  differential thermal analysis, 88
  dust explosions, 90
  flammable limits, 90
  flash point, 90
  shock sensitivity, 89
  testing, 87–90
Reactive hazard evaluations, 86
Renormalization-group theories, 148–151
  excluded-volume field, 149
  free-energy density, 149–150

Renormalization-group theories *(continued)*
  osmotic pressure, 150–151
  scaling, 164–169
  semidilute solutions, 149
Resista-Clad, 107–108
Rheological models of suspensions, 1–67, *see also* Momentum tracer methods; Spatially periodic suspensions
  antisymmetric stress, 66
  averaged continuity equation, 29
  averaging Stokes equations, 29
  bond percolation problems, 32–33
  Brinkman equation, 29–30
  Brownian motion, 4
  bulk conductance, 33
  cell models, 5, 21–22
  clusters, 35
  coefficient k_1, 23–24
  complicating effects, 4
  continuum-mechanical principle of material-frame indifference, 67
  contribution of Brownian motion, 26
  critical concentration, 35–36
  dilute suspensions, 23–27
  empirical models, 19–21
  evolution equation, 54–55
  Fickian diffusion tensors, 26–27
  fixed particles, 30
  force-torque vector, 54
  fractal suspensions, 64–66
  infinite viscosity prediction, 19–20
  inner zone dissipation, 22
  internal spin field, 66
  Leibniz packing, 64–65
  Liouville equation, 25
  long-time mean structure, 56
  mechanical energy dissipation, 23
  modified Menger sponge, 65–66
  momentum transport, 34
  Monte Carlo techniques, 31
  notation and scope, 5–6
  particle-particle interactions, 23–24
  percolation models, 32–36
    application, 34–36
    theory, 32–34
  percolation-theory approaches, 5
  phase function, 28–29
  relative viscosity, 22, 56
  repulsive interparticle forces, 57
  sedimentation velocity, 26–27
  shear-induced collisions, 34
  shearing in macro-couette apparatus, 35
  spatially periodic model, 35–36
  statistical models
    formal expansions, 28–31
    numerical calculations, 31–32
  Stokesian dynamics, 54–57
  Stokes problem, 25
  suspension of spheres, 30
  three-phase systems, 22
  volume fraction, 21
  vortex viscosity, 66–67
  wave vector-dependent viscosity, 31
Rheology
  microrheology of suspensions, 20
  polymer–colloid–solvent mixtures, 219–224
  suspensions, 44–47
Root–mean–square layer thickness, 173

S

Scaling, renormalization group theories, 164–169
Scaling theory, 144–148
  correlation length, 145–147
  crossover from dilute to semidilute, 145–146
  density fluctuations, 145
  osmotic pressure, 146–148
Scheutjens–Fleer lattice model, 169–179
  adsorbed amount, 171
  adsorbed layer structure, 172–174
  adsorbed polystyrene, 178
  adsorption isotherms, 173–174
  chemical potential, 176
  conclusions, 179
  depth of attractive minimum, 176–177
  force measurements, 176–178
  Helmholtz free energy, 175–176
  interaction potential energy, 176–178
  interactions between layers, 174–178
  lattice sites, 170
  layer thickness, 173–175
  motivation and formulation, 169–172
  nonlocal segment–solvent interactions, 171
  segment distribution, 172–173
  volume-fraction profiles, 172
Seal-less pumps, 128–129
Sedimentation velocity, 26–27
Self-consistent field theory, 179–197
  accuracy, 196–197

adsorbed amount versus chain length, 188–189
adsorption isotherms, 188
chemical potential, 193
configuration integral, 186–187
continuous and discontinuous surface functions, 180
depth of attractive minima, 194–195
Drude equations, 189–190
ellipsometric thickness as function of chain length, 190–191
equation-of-state, 193
Flory–Huggins equation-of-state, 180–181
ground state solutions, 183–184
Helmholtz free energy, 191
hydrodynamic thickness, 191–192
interaction potential energy versus gap half-width, 193–196
interactions between layers, 191–195
isolated layers, 187–191
Laplace transforms, 185
matched asymptotic expansion, 184–187
mixed boundary condition, 181–182
motivation and formulation, 179–183
osmotic pressure, 192
Ploehn–Russel model, 196
sticky surface model, 180
volume fraction profiles, 187–188
Self-consistent, mean-field theories, 153–156
configuration integral, 154
equilibrium spatial distribution, 156
Helmholtz free energy, 155–156
Shear-induced particle diffusion, 20
Shock sensitivity, reactive chemicals, 89
Shutdown, large continuous processes, 112–113
Silica, apparent viscosity of dispersions, 220–222
Single mechanical seals, 129–130
Sleeve-type plug-cocks, 123
Spatially periodic model, 35–36
Spatially periodic suspensions, 36–53
adherence condition, 52
centrosymmetric particles, 45
coefficients, 49
collision sphere, 40
cubic arrays, 48–51
description, 38–39
dilute arrays, 49
ergodic character of process, 47, 53
extension to N particles in unit cell, 51–53
extrapolation to infinite suspension, 53
force/torque balance, 52
generic formula, 46
grand resistance matrix, 45, 51–52
interaction of vector with unit square, 43–44
kinematical problems, 36–37
kinematics, 38–44
latice configuration, 42–43
latice deformation, 39
macroscopic velocity gradient dyadic, 39
maximum density, 41–42
nonstatic, sheared systems, 40
no-slip boundary condition, 44
perfect crystal problem, 37
rheology, 44–47
self-coincidence symmetry operations, 38–39
simple shear flow, 42–43
star body, 41
Stokes equations, 45, 48
time dependence of particle configuration, 50
touching-sphere limit, 50
trajectory equation, 52
translational velocity, 44
two-dimensional, incompressible linear flow, 40
velocity field, 44
Spiral wound gaskets, 121–122
Spring-loaded check valves, 124
Static mixer reactors, 93
Sticky surface model, 180
Stokes equations
averaging, 29
momentum tracer methods, 59–60
quasistatic, 7
spatially periodic suspensions, 45, 48
Stokesian dynamics, 54–57
Stokes problem, 25
Stress coefficients, 56
Stress crack corrosion, metals, 105–106
Stress tensor, suspensions, 8
Suspensions, *see also* Rheological models of suspensions
basic equations and properties, 6–10
conditional probability density, 15
configuration-dependent tensorial quantity, 15
continuity equation, 7
homogeneous, volume averages, 16–17

hydrodynamic force, 8
incorporation of Brownian motion, 15
Liouville equation, 15
macroscopic stress, 17
macroscopic velocity gradient, 16
many-body problem, low Reynolds-number hydrodynamics, 10–13
method of induced forces, 12
microrheology, 20
particle-specific probability densities, 14
particle stress, 8
partitioned matrix relation, 9
position-independent velocity gradient, 6
quasistatic Stokes equation, 7
shear-induced particle diffusion, 20
statistical formulation, multiparticle problem, 13–15
stress coefficients, 56
stress tensor, 8
torque, 8
two-body methods, 10–12
two-sphere systems, 10
velocity field, 7–8
volumetric particulate number density, 14
Suspension theories, major reviews, 2–3
Styrene-butadiene latex, preparation, 91

T

Tandem-seal pumps, 131–132
Teflon, 109–110
  envelope gaskets, 123
Tensor fields, momentum tracer methods, 61
Torque, suspensions, 8

Total containment, chemical plant design, 83–84
Trajectory equation, 52
Transparent devices, 126–127
Transport equation, momentum tracer methods, 61
Triblock copolymer, 138–139

U

Underground storage tanks, standards, 82
Unit cell equations, steady-state, 61

V

Vacuum relief systems, 126
Valves, chemical plant design, 123–124
Venting, dust explosions, 97–98
Vent panels, 98
Viscosity
  apparent, adsorbing homopolymer, 220–222
  kinematic, 58–60
  relative, 56
  steady shear, polystyrene latices, 222–223
  vortex, magnetic fluids, 66–67
  wave vector-dependent, 31
Vortex viscosity, magnetic fluids, 66–67

W

Worst case thinking, chemical plant design, 86–87